Circular Economy Practices in Solid Waste Value Chains

This book presents a detailed analysis of the potential of the circular economy (CE) to revolutionize solid waste management (SWM) in developing economies. It bridges the gap between theory and practice by offering real-world, actionable insights and examples to support the development of sustainable and resilient SWM systems. It focuses on developing economies across several regions, including (but not limited to) sub-Saharan Africa, South and Southeast Asia, Latin America and the Caribbean, and the Middle East and North Africa, and includes relevant case studies.

Key features include:

- Detailed coverage of the solid waste value chain
- Strong emphasis on practical circular economy practices
- Review of measures for implementing practical strategies
- Focus on perspectives from developing economies
- Use of case studies to demonstrate circular economy practices in developing contexts.

This book is intended for students, researchers, policymakers, practitioners, and community leaders working in the field of environmental engineering.

Africa Circular Economy Series

CRC Press (Taylor & Francis Group) introduces **Africa Circular Economy Series** (with a Science and Technology outlook but preferably more interdisciplinary). Under this initiative, we invite scholars, academicians, researchers, and professionals to contribute to this series. We are interested in discussing monographs, references, textbooks, shortform books, and handbooks to add to our book programs, at undergraduate, postgraduate, and doctoral levels. All the books published under the series will be part of our Global Publishing program.

Circular and Transformative Economy
Advances towards Sustainable Socio-economic Transformation
Edited by Luxon Nhamo, Sylvester Mpandeli, Stanley Liphadzi and Tafadzwanashe Mabhaudhi

Municipal Solid, Agricultural, and Mining Waste in Sub-Saharan Africa
Sustainable and Circular Valorization Opportunities
Freeman Ntuli, Tebogo Mashifana, Thandiwe Sithole and Thabo Falayi

Computational Fluid Dynamics Modelling of Solid Fuel Co-Combustion
Garikai T. Marangwanda, Daniel M. Madyira, and Innocent Musonda

Circular Economy Practices in Solid Waste Value Chains
A Developing Economies Perspective
Bupe Getrude Mutono-Mwanza

For more information about this series, please visit: www.routledge.com/Africa-Circular-Economy-Series/book-series/CRCACES

Circular Economy Practices in Solid Waste Value Chains

A Developing Economies Perspective

Bupe Getrude Mutono-Mwanza

CRC Press
Taylor & Francis Group
Boca Raton London New York

CRC Press is an imprint of the
Taylor & Francis Group, an **informa** business

Designed cover image: Bupe Getrude Mutono-Mwanza

First edition published 2027
by CRC Press
2385 NW Executive Center Drive, Suite 320, Boca Raton FL 33431

and by CRC Press
4 Park Square, Milton Park, Abingdon, Oxon, OX14 4RN

CRC Press is an imprint of Taylor & Francis Group, LLC

ISBN: 9781032958910 (hbk)
ISBN: 9781032970288 (pbk)
ISBN: 9781003591764 (ebk)

DOI: 10.1201/9781003591764

Typeset in Times
by codeMantra

This book is dedicated to those devoted to promoting the circular economy (CE), particularly in relation to solid waste management (SWM) in developing economies, and to the collective insights gained from the experiences of developed economies. It recognizes scholars, policymakers, practitioners, entrepreneurs, and communities who operate in diverse contexts to convert waste into value, enhance inclusive systems, and close global knowledge gaps. May this endeavor stimulate evidence-based learning, adaptive innovation, and collaborative efforts that foster sustainable, resilient, and socially equitable circular transitions in developing economies.

Contents

About the Author xiii
Preface........ xv
Acknowledgments........ xvii
List of Abbreviations........ xix

Chapter 1 Understanding Circular Economy........ 1

1.1 Definition of Circular Economy........ 1
1.2 Significance of CE in Solid Waste Management........ 2
1.2.1 Waste Reduction and Resource Efficiency........ 3
1.2.2 Economic Opportunities and Job Creation........ 3
1.2.3 Environmental Sustainability and Climate Change Mitigation........ 4
1.2.4 Support for Informal Sector Integration........ 4
1.2.5 Policy Innovation and Sustainable Urban Development........ 5
1.3 Global Solid Waste Challenges and Sustainable Interventions........ 5
1.3.1 Growing Volumes of Municipal Solid Waste........ 6
1.3.2 Inequality in Waste Management Capacity........ 7
1.3.3 Environmental and Public Health Impacts........ 7
1.3.4 Emerging and Complex Waste Streams........ 7
1.3.5 Socioeconomic Dimensions of Waste........ 8
1.3.6 The Need for Sustainable Practices........ 8
1.3.7 Responding to Growing Waste Volumes........ 9
1.3.8 Reducing Inequalities in Waste Management Capacity........ 9
1.3.9 Mitigating Environmental and Public Health Risks........ 9
1.3.10 Addressing Emerging and Complex Waste Streams........ 10
1.3.11 Tackling Socioeconomic Vulnerabilities........ 10
1.3.12 Aligning with Global Sustainability Goals........ 10
1.4 Developing Economies........ 11
1.4.1 Facets of Developing Economies........ 11
1.4.2 Economic Structure........ 12
1.4.3 Demographics........ 12
1.4.4 Institutions and Governance........ 13
1.4.5 Infrastructure and Technology........ 14
1.4.6 Social and Environmental Dimensions........ 15
1.5 SWM in Developing Economies........ 15
1.5.1 High Waste Generation with Limited Collection........ 16

1.5.2 Dominance of Organic Waste 17
1.5.3 Informal Sector Involvement 17
1.5.4 Limited Financial Resources 18
1.5.5 Environmental and Health Risks 18
1.6 Socioeconomic and Environmental Effects of Mediocre Waste Management 18
References 20

Chapter 2 The Solid Waste Value Chain 26

2.1 Generation of Solid Waste 26
2.1.1 Sources of Solid Waste 26
2.1.2 Types of Solid Waste 26
2.1.3 Characterization of Solid Waste in Developing Economies 30
2.2 Collection and Transportation 31
2.2.1 Methods of Waste Collection 31
2.3 Transportation Models and Challenges 33
2.3.1 Direct Haul Model 33
2.3.2 Station or Transfer-Based Model 33
2.3.3 Decentralized/Community-Based Model 33
2.3.4 Private Sector and Informal Model 34
2.3.5 Intermodal Model 34
2.4 Challenges in Waste Transportation 35
2.4.1 Infrastructure Limitations 35
2.4.2 High Operational Costs 35
2.4.3 Limited Transfer Stations 35
2.4.4 Weak Integration of Informal Sector 36
2.4.5 Environmental and Health Concerns 36
2.4.6 Institutional and Financial Barriers 36
2.4.7 Urbanization and Increasing Waste Volumes 37
2.5 Role of Informal Sector in Waste Collection 37
2.5.1 Contribution to Waste Collection and Recycling 38
2.5.2 Livelihoods and Socioeconomic Significance 38
2.5.3 Environmental Impact 38
2.5.4 Challenges Facing the Informal Sector 39
2.5.5 Pathways for Integration and Formalization 39
2.6 Processing and Treatment 40
2.6.1 Methods for Waste Segregation and Processing 40
2.7 Technologies for Waste Treatment 43
2.7.1 Composting 43
2.7.2 Anaerobic Digestion 44
2.7.3 Waste-to-Energy 44
2.7.4 Recycling and Materials Recovery Facilities 45
2.7.5 Digital and Integrated Technologies 45
2.7.6 Comparative Analysis 46

2.8 Case Studies from Developing Economies ... 46
2.8.1 India—Decentralized Composting Anchored by Source Segregation ... 46
2.8.2 Ethiopia and South Africa—Anaerobic Digestion Linking Waste, Energy, and Fertility ... 47
2.8.3 India—Waste-to-Energy under Mixed, Moist Waste Conditions ... 47
2.8.4 Johannesburg, South Africa—Hybrid Recycling with MRFs and Informal Integration ... 48
2.8.5 Southeast Asia and Sub-Saharan Africa—Digital/Industry 4.0 Pilots for Integrated Operations ... 48
2.9 Recycling and Recovery ... 49
2.9.1 Significance of Recycling in the CE ... 49
2.9.2 Recycling Systems and Processes ... 51
2.9.3 Market for Recycled Materials ... 53
2.10 Disposal and Landfilling ... 55
2.10.1 Traditional Disposal Methods and Their Demerits ... 55
2.10.2 Modern Landfilling Techniques ... 56
2.10.3 Health and Environmental Effects of Illegal Solid Waste Disposals ... 57
References ... 58

Chapter 3 Implementing Circular Economy in Solid Waste Management ... 62

3.1 Policy and Regulatory Frameworks ... 62
3.1.1 Existing Regulations and Policies in Developing Economies ... 62
3.1.2 Best Practices from Developed Economies ... 64
3.2 Financial and Economic Instruments ... 69
3.2.1 Investments Opportunities and Funding Mechanisms ... 70
3.2.2 Economic Incentives for Consumers and Businesses ... 71
3.3 Technological Innovations ... 72
3.3.1 Emerging Technologies in SWM ... 73
3.3.2 Digital Solutions for SWM and Tracking ... 74
3.3.3 Technology Adaptation for Developing Economies ... 76
3.4 Community Engagement and Awareness ... 77
3.4.1 Strategies for Community Engagement and Awareness ... 77
3.4.2 Strategies for Influencing Public Participation ... 80
3.4.3 Strategies for Changing Behaviors ... 82
3.5 Case Studies of Successful Community Initiatives ... 84

3.5.1 Community-Based Circular Waste Management in Tan Deaw Sub-district, Thailand ... 84
3.5.2 Household Participation in Chelstone Township, Lusaka, Zambia ... 85
3.5.3 Reward Recycling and Pay-As-You-Throw in Northern Greece ... 85
References ... 85

Chapter 4 Sector-Specific Applications ... 90

4.1 Urban Solid Waste Management ... 90
4.1.1 Challenges of Urban Waste Management in Developing Cities ... 90
4.1.2 Innovative Urban Waste Management Solutions ... 91
4.2 Case Studies from Major Developing Cities ... 94
4.2.1 Urban Symbiosis Model: Lessons for Developing Economies ... 94
4.2.2 Case of Ahmedabad, India: Social Innovation for Circular Waste Management ... 94
4.2.3 Case Study: Smart City Circular Waste Innovations in Nigeria ... 95
4.2.4 Case Study: Circular and Community-Based Waste Innovation in Mombasa County, Kenya ... 96
4.3 Rural Solid Waste Management ... 96
4.3.1 Unique Challenges of Rural Waste Management ... 97
4.3.2 Tailored Solutions for Rural Areas ... 99
4.3.3 Successful Rural Waste Management Programs ... 101
4.4 Industrial Solid Waste Management ... 102
4.4.1 Handling Industrial and Hazardous Waste ... 103
4.4.2 CE Practices for Industrial Waste ... 104
4.4.3 Case Studies of Industrial Waste Management ... 105
4.5 Agricultural and Organic Waste Management ... 107
4.5.1 Importance of Managing Agricultural Waste ... 107
4.5.2 Circular Practices for Agricultural Waste ... 108
References ... 111

Chapter 5 Case Studies and Best Practices ... 116

5.1 Successful Implementations ... 116
5.1.1 Implementation of CE in a Developing Economy's Mining Industry Using Institutional Theory: The Case of Namibia ... 116
5.1.2 CE Business Models in Developing Economies: Lessons from India ... 117

5.1.3 Implementation of CE Principles in Industrial SWM: Case Studies from a Developing Economy 118
5.1.4 CE Model for Developing Countries: Evidence from Bangladesh 119
5.2 Challenges and Failures in Implementation 120
5.2.1 Barriers and Challenges to Waste Management Hindering the CE in Sub-Saharan Africa 120
5.2.2 Challenges of the South African Economy to Transition to a CE: A Case of Remanufacturing 121
5.2.3 Barriers Toward CE Transition: Exploring Different Stakeholders' Perspectives 122
5.2.4 Implementing CE Practices in Indian SMEs: Analysis of Challenges with Case Study 123
References 125

Chapter 6 The Future of Circular Economy in Solid Waste Management 126

6.1 Overview 126
6.2 Trends and Emerging Opportunities 126
6.2.1 Future Trends in SWM and CE 126
6.3 Emerging Opportunities for Developing Economies 128
6.3.1 Entrepreneurship and Innovation as Engines of CE Growth 128
6.3.2 Digital Platforms and Technology-Enabled Circular Opportunities 128
6.3.3 Sector-Specific Opportunities Across Developing Economies 128
6.3.4 Collaboration and Networks as Enablers of Circular Value 129
6.3.5 Systems Thinking and Integrated Circular Transitions 129
6.3.6 Employment, Inclusion, and Social Value Creation 129
6.4 Role of Research and Innovation 130
References 131

Chapter 7 Roadmap for Developing Economies 132

7.1 Overview 132
7.1.1 Pillar 1: Policy and Institutional Alignment 133
7.1.2 Pillar 2: Circular Business Models and Entrepreneurship 133

7.1.3 Pillar 3: Digitalization, Research, and Innovation 134
7.1.4 Pillar 4: Social Inclusion and Skills Development 134
7.2 Integrated Four-Pillar Framework 135
7.3 Key Stakeholders and Their Roles 137
7.3.1 Government and Public Institutions 137
7.3.2 Private Sector and Industry 138
7.3.3 Communities, Informal Sector, and Consumers 138
7.3.4 Research Institutions and Innovation Systems 139
References 139

Chapter 8 Conclusion 141

8.1 Overview 141
8.2 Core Insights on CE Thinking 141
8.3 Significance of CE in SWM 142
8.3.1 Moving Beyond Collection and Disposal 142
8.3.2 Waste as a Resource and Economic Opportunity 142
8.3.3 Governance, Policy, and Institutional Alignment 143
8.3.4 Business Models, Entrepreneurship, and Supply Chains 143
8.3.5 Digitalization, Data, and Innovation in Waste Systems 143
8.3.6 Social Inclusion, Informality, and Skills Development 144
8.4 Implications for Sustainable Development 144
8.5 Forward-Looking Reflections and Strategic Direction 145
Bibliography 145

Index 147

About the Author

Bupe Getrude Mutono-Mwanza is Associate Professor in Sustainable Supply Chains and Circular Economy and Associate Director at the Graduate School of Business, University of Zambia. She is also a Professor Extraordinarius at the University of South Africa. With over a decade of academic leadership, research, postgraduate supervision, and teaching experience, she is a recognized scholar in sustainable supply chains and circular economy transitions. She holds a BSc in Production Management, an MEng in Manufacturing Systems and Operations Management, and a PhD in Engineering Management. She has published widely in international journals, authored a book on the circular economy and sustainable operations, and presented award-winning papers at global conferences. Her research focuses on sustainable operations, resource recovery, and waste valorization, making significant contributions to Zambia's scholarship on circular value chains. She also serves in professional and governance roles and actively engages in public scholarships, promoting sustainability and inclusive development.

Preface

Circular Economy Practices in Solid Waste Value Chains provides a comprehensive yet practical roadmap for transforming solid waste management systems in developing economies through circular economy principles. This book combines theoretical foundations with applied strategies, examining the full solid waste value chain from generation and collection to recycling, recovery, and reintegration into production systems. It draws on experiences from sub-Saharan Africa, South and Southeast Asia, Latin America, and the Middle East, offering context-sensitive insights rather than transplanting models from high-income countries.

Key features include analysis of policy frameworks, circular business models, digital innovation, social inclusion, and sector-specific applications across multiple waste streams. This book integrates case studies, best practices, and actionable implementation strategies to support evidence-based decision-making.

The primary audience includes students, researchers, policymakers, municipal authorities, development practitioners, industry leaders, and community organizations engaged in sustainability and waste governance. Readers will benefit from a structured framework for designing inclusive and economically viable circular transitions, gaining tools to align environmental objectives with competitiveness, job creation, and resilience. Ultimately, this book supports practical reform of solid waste systems while positioning the circular economy as a developmental pathway for emerging economies.

Acknowledgments

I devote all glory, honor, and gratitude to God Almighty, whose wisdom, grace, and unwavering guidance have made the completion of this book possible. Throughout each phase of conceptualization, investigation, writing, and reflection, His strength has supported me, and His purpose has imbued this work with significance. Without His divine guidance and provision, this voyage would not have been accomplished.

My heartfelt appreciation is extended to my family, whose love, patience, encouragement, and steadfast support sustained me throughout the extensive hours of writing and revision. Their comprehension, sacrifices, and conviction in the significance of this work offered both inspiration and equilibrium, emphasizing the importance of service, perseverance, and humility.

I genuinely appreciate the numerous scholars, practitioners, policymakers, and organizations whose research, articles, reports, and case studies have informed and enriched this publication. The authors of the diverse academic publications and real-world case studies cited herein have provided essential knowledge and insights that have influenced the perspectives and arguments articulated presented in this book. Their dedication to promoting circular economy (CE) principles and sustainable solid waste management (SSWM) across various contexts has been fundamental to the development of this publication.

Finally, I extend my recognition to all practitioners and communities actively engaged in solid waste management (SWM) and CE initiatives. Their experiences, innovations, and resilience continue to inspire pathways toward more sustainable, inclusive, and resilient systems.

Abbreviations

8NDP	Eighth National Development Plan
AD	Anaerobic Digestion
AfDB	African Development Bank
AI	Artificial Intelligence
CE	circular economy
CSR	Corporate Social Responsibility
DBSA	Development Bank of Southern Africa
DRS	Deposit Return Scheme
EPR	Extended Producer Responsibility
ESG	Environmental, Social and Governance
E-waste	Electronic Waste
GCE	Green Circular Economy
GEF	Global Environment Facility
GESIP	Green Economy and Sustainable Industrialization Programme
GHGs	greenhouse gases
GIS	Geographic Information System
IoT	Internet of Things
IS	Industrial Symbiosis
LCA	Life Cycle Assessment
MBT	Mechanical Biological Treatment
MRFs	Materials Recovery Facilities
MSW	Municipal Solid Waste
NGO	Non-Governmental Organization
OECD	Organization for Economic Co-operation and Development
PAR	Participatory Action Research
PAYT	Pay-As-You-Throw
PET	Polyethylene Terephthalate
PP	Polypropylene
RDF	Refuse-Derived Fuel
SDGs	Sustainable Development Goals
SSWM	Sustainable Solid Waste Management
SWM	Solid Waste Management
UN	United Nations
UNEP	United Nations Environmental Protection
W-t-E	waste-to-energy

1 Understanding Circular Economy

1.1 DEFINITION OF CIRCULAR ECONOMY

CE is a developing economic concept aimed at converting the conventional "take-make-dispose" linear economy into a regenerative system that maximizes resource utilization over extended periods. It fundamentally focuses on eliminating waste and pollution, maintaining the circulation of products and materials, and restoring natural systems (Ellen MacArthur Foundation, 2013). This transcends traditional recycling, representing a holistic methodology for the design, production, consumption, and recovery of products. To obtain a clear and agreeable definition of CE, many authors have conducted studies centered on defining CE. For example, Kirchherr et al. (2017) define CE as an economic system that is based on business models which replace the "end-of-life" concept with reducing, alternatively reusing, recycling and recovering materials in production/distribution and consumption processes, thus operating at the micro level (products, companies, consumers), meso level (eco-industrial parks), and macro level (city, region, nation, global), with the aim to accomplish sustainable development, which implies creating environmental quality, economic prosperity, and social equity, to the benefit of current and future generations.

In another study by Kirchherr et al. (2023), CE is redefined as an economic framework that promotes sustainable production and consumption by decelerating, closing, and constricting material and energy cycles. This approach aims to minimize resource extraction, waste, emissions, and energy loss while encouraging systemic innovation at micro (products, companies, consumers), meso (eco-industrial parks), and macro (cities, regions, nations, global) levels. The objective is to achieve sustainable development that harmonizes environmental quality, economic prosperity, and social equity for the benefit of present and future generations.

Figge et al. (2023) offer a nuanced and dynamic definition of CE as a holistic strategy for resource management aimed at slowing, reducing, and eradicating material and energy cycles, with the primary goal of achieving ecological sustainability, rather than merely economic efficiency or resource conservation. This concept is particularly relevant to discussions in closing the loop, since it aligns with this book's claim that CE strategies in solid waste value chains must achieve environmental and social goals alongside economic ones.

Authors such as Geisendorf and Pietrulla (2018) define CE as an economic system which replaces the concept of end of life with reducing, reusing, recycling, and recovering materials throughout the production, distribution, and consumption processes. It operates at the micro level (products, companies, consumers), meso level (eco-industrial parks), and macro level (city, region, nation, global), with the

DOI: 10.1201/9781003591764-1

aim to achieve sustainable development, creating environmental quality, economic prosperity, and social equity for the benefit of current and future generations.

Lastly, Awan et al. (2020) define CE as an economic model that prioritizes the sustainable utilization of resources by preserving the value of products, materials, and resources within the economy for an extended duration, while minimizing waste through strategies such as reduction, reuse, recycling, remanufacturing, and recovery, ultimately aiming to promote sustainable development.

The notion of the CE has markedly evolved in literature, with varying emphases among authors. Kirchherr et al. (2017) offered a prominent definition, characterizing the CE as an economic system that substitutes the "end-of-life" paradigm with strategies of reduction, reuse, recycling, and recovery at micro, meso, and macro levels, with sustainable development as the overarching objective. This definition, although thorough, has been criticized for its excessive breadth and significant overlap with sustainability discussions. Geisendorf and Pietrulla (2018) emphasized the systemic nature of CE and its triple-bottom-line focus, although contributed minimal innovation beyond Kirchherr's framework. Awan et al. (2020) enhanced the discourse by characterizing CE as a paradigm shift that prioritizes value retention and resource efficiency, particularly in logistics and supply chain contexts, despite their approach being predominantly abstract and focused on the Global North. Recently, Kirchherr et al. (2023) refined the notion by explicitly integrating the mechanics of slowing, shutting, and shrinking resource loops, framing the CE as a transition pathway necessitating systemic innovation. Figge et al. (2023) enhanced operational clarity by delineating CE in terms of minimizing material and energy throughput, emphasizing specific techniques like design, repair, reuse, remanufacturing, and composting. However, their definition constricts CE to circularity mechanics, minimizing the significance of social fairness and justice aspects.

Collectively, these definitions demonstrate a conflict between expansive sustainability frameworks and more focused loop-based implementations. A significant deficiency across all areas is the insufficient focus on the realities of developing/emerging economies, where informal sectors, inadequate regulatory frameworks, poverty alleviation, and livelihood improvement are essential to circular practices. From this viewpoint, the CE in developing economies can be defined as:

> "a socio-economic framework aimed at reducing material and energy consumption by decelerating, closing, and constricting resource cycles through repair, reuse, recycling, remanufacturing, and recovery." It amalgamates both formal and informal sectors to preserve value in products, materials, and resources for an extended duration, while concurrently improving livelihoods, alleviating poverty, and promoting environmental resilience. The primary objective is sustainable development encompassing ecological preservation, economic efficiency, social fairness, inclusiveness, and equitable transitions for present and future generations.

1.2 SIGNIFICANCE OF CE IN SOLID WASTE MANAGEMENT

CE has arisen as a revolutionary strategy to tackle the ongoing issues of SWM in developing economies. In contrast to the conventional linear model of "take–make–dispose," CE prioritizes resource efficiency, reuse, recycling, and recovery, resulting in environmental and socioeconomic advantages (Kirchherr et al., 2017). Global empirical evidence indicates that CE-based waste treatment methods can markedly diminish waste footprints while improving sustainability results (Tisserant et al., 2017). From the standpoint of a developing economy, CE offers context-specific remedies to issues including insufficient infrastructure, financial limitations, and dependence on informal waste systems by fostering resource recovery and inclusive value chains (Mandpe et al., 2023). Furthermore, theoretical and practical insights demonstrate the implementation of CE economy principles through strategies like extended producer responsibility (EPR), waste segregation, and eco-industrial parks, thereby converting sustainability concepts into concrete waste management practices (Dumlao-Tan & Halog, 2017). The studies above do demonstrate the significance of CE in SWM, but it is necessary to discuss further other factors that illustrate the significance of CE in SWM from a developing economy perspective, and in this book, it is necessary that these factors are demonstrated in detail.

1.2.1 Waste Reduction and Resource Efficiency

Developing economies frequently encounter restricted landfill capacity and insufficient waste infrastructure, resulting in open dumping and incineration of waste, hence posing environmental and public health hazards. CE activities, including material recovery, composting, and recycling, substantially diminish the quantity of waste directed to landfills, while enhancing resource efficiency. This is especially significant in areas where urbanization and population expansion are increasing garbage production (World Bank, 2018). In addition to landfill diversion, CE strategies provide energy recovery from waste, hence enhancing sustainability and energy security in resource-limited economies (Elroi et al., 2023). Moreover, enhancing waste resource efficiency creates economic value by prolonging material lifecycles and diminishing dependence on virgin resources, but governmental and institutional obstacles must be overcome to fully actualize these advantages (Van Ewijk, 2018). To close loop in solid waste value chains, it is important that the stakeholders in the value chains understand that CE does enhance waste reduction and resource efficiency, and this is possible in developing economies.

1.2.2 Economic Opportunities and Job Creation

CE-based SWM generates novel economic prospects in recycling, repair, remanufacturing, and waste valorization sectors. Waste pickers in African and Asian cities are integral to recycling value chains; formalizing their roles through CE models can improve livelihoods, encourage entrepreneurship, and stimulate inclusive economic growth (Wilson et al., 2006; Azevedo et al., 2019). Evidence from Europe indicates that transitions to a CE in waste management can not only provide additional employment opportunities but also enhance job quality, including increased salaries and working conditions (Weghmann, 2017). In rapidly urbanizing environments like Indonesia, the incorporation of CE into waste management has created entrepreneurial prospects, allowing small enterprises to flourish while tackling urban garbage issues (Wikurendra et al., 2024). Furthermore, CE-driven employment generation corresponds with the UN Sustainable Development Goals by promoting both equitable labor and sustainable urban development (Schroeder et al., 2019). These studies demonstrate why developing countries should drive the agenda of closing the loop in solid waste values chains because of the opportunities that CE presents.

1.2.3 Environmental Sustainability and Climate Change Mitigation

CE techniques mitigate greenhouse gas (GHG) emissions by redirecting organic waste to composting or biogas generation and by reducing methane emissions from uncontrolled landfills. This corresponds with the climate obligations of developing nations as stipulated in the Paris Agreement. The UNEP (2021) indicates that enhancing circular methods in waste management might diminish world emissions by as much as 20% by 2050. Empirical research indicates that circular solutions, including waste-to-energy and recycling, are essential strategies for addressing climate change and other environmental challenges (Yang et al., 2025). Case studies from Brazil illustrate that CE-based waste management in urban settings not only decreases greenhouse gas emissions but also improves resilience and resource efficiency (Paes et al., 2024; Bellezoni et al., 2022). In addition to municipal solid waste (MSW), CE techniques in the built environment, including material reuse and low-carbon construction, demonstrate significant mitigation potential (Gallego-Schmid et al., 2020). CE is becoming acknowledged as a strategic method for reducing emissions and ensuring long-term environmental sustainability in emerging economies (Hailemariam & Erdiaw-Kwasie, 2023). Therefore, mitigating climate change and achieving environmental sustainability is supported by discussions from the authors above, and this book presents how this can be achieved.

1.2.4 Support for Informal Sector Integration

In numerous emerging economies, the informal sector constitutes a substantial portion of waste collection and recycling activities. CE establishes a framework for incorporating these stakeholders into official waste management systems, guaranteeing equitable compensation, enhanced working conditions, and social acknowledgment, while simultaneously increasing overall recycling rates (Gutberlet & Uddin, 2017). Data from India indicates that integrating informal and formal sectors in e-waste management within a CE framework not only improves material recovery but also mitigates environmental hazards (Sengupta et al., 2022; Mehra et al., 2026). Research indicates that the combination of formal and informal systems enhances the economic impact of waste pickers while promoting efficiency and sustainability in developing contexts (Majeed & Luni, 2020). The informal recycling industry is increasingly seen as vital for promoting the Sustainable Development Goals, especially those pertaining to decent labor and responsible consumerism (Valencia et al., 2022). Recent research underscores that formalizing the informal e-waste sector is a complicated yet essential step in advancing towards a CE (Pandya, 2024) in developing economies.

1.2.5 Policy Innovation and Sustainable Urban Development

By embracing CE ideas, municipalities in emerging economies can transition to integrated SWM (ISWM), which merges environmental conservation with socio-economic advancement. This facilitates the attainment of the UN Sustainable Development Goals (SDGs), namely Goal 11 (sustainable cities), Goal 12 (responsible consumption and production), and Goal 13 (climate action) (Schroeder et al., 2019). Data from developing economies indicates that CE-oriented SWM offers context-specific strategies for sustainable growth, harmonizing environmental sustainability with economic prospects (Mandpe et al., 2023). In Brazil, urban CE evaluation frameworks have been utilized to evaluate and enhance the transition to circular cities, demonstrating how municipalities can implement CE principles in practice (Moraes et al., 2023). Likewise, CE and waste management are progressively acknowledged as strategic instruments for constructing climate-neutral urban futures, placing municipal governance at the center of climate initiatives (Möslinger et al., 2023). In addition to urban settings, CE has prospects for sustainable rural development, wherein community-driven waste valorization fosters inclusive and regionally equitable sustainability transitions (Mihai, 2023).

1.3 GLOBAL SOLID WASTE CHALLENGES AND SUSTAINABLE INTERVENTIONS

SWM is increasingly seen as one of the most pressing sustainability concerns of the 21st century. Increasing populations, swift urbanization, industrial growth, and changing consumer patterns are causing unparalleled levels of waste production worldwide. The World Bank (2018) reported that worldwide solid waste production was 2.01 billion tons per year in 2016 and is anticipated to rise to 3.40 billion tons by 2050 if current trends continue. Although waste creation per capita is greatest in high-income nations, the most significant increases are anticipated in developing economies, where urbanization is rapidly advancing and waste management infrastructure frequently falls short. The expanding disparity underscores the dual difficulty of addressing increasing waste volumes and inherent inequities in waste governance.

Research highlights that the issue is especially severe in the Global South, where constrained financial resources, fragile institutional frameworks, and insufficient infrastructure exacerbate the effects of increasing waste (Abubakar et al., 2022). Waste management in numerous developing nations is persistently underfunded, resulting in extensive dependence on open dumping, backyard incineration, and unregulated landfilling. These practices result in substantial environmental and health repercussions, including groundwater pollution and respiratory diseases, while also markedly increasing greenhouse gas emissions. Kabir and Kabir (2021) noted over a decade ago that these systemic hurdles would present future concerns, and current evidence substantiates that these issues have indeed escalated.

The dilemma is not limited to developing economies alone. Wilson and Velis (2017) contend that SWM constitutes a worldwide catastrophe necessitating immediate and coordinated evidence-based intervention. They emphasize that no nation, regardless of its development status, has completely addressed the intricacies of contemporary waste management systems. The escalating magnitude and variety of waste streams, along with insufficient control, render waste a crucial environmental challenge of our era. Chen et al. (2020) offer empirical evidence indicating that the growth trends of MSW are accelerating worldwide due to demographic shifts, increased affluence, and changing lifestyles. The environmental and socioeconomic ramifications of this trend surpass waste disposal, affecting climate change, public health, and sustainable urban growth.

A further aspect of the challenge is the rise of intricate waste streams. Conventional systems designed for the management of organic and municipal waste are being inundated by electronic waste, plastics, hazardous substances, and building debris. The newly established waste categories necessitate sophisticated treatment, oversight, and regulatory frameworks, which are absent in numerous nations. As a result, the disparity between the volume of waste produced and the ability of institutions to manage it effectively continues to expand.

Considering this setting, experts have increasingly emphasized the necessity for a paradigm shift in waste governance. Linear, end-of-pipe methodologies that include the collection and disposal of waste with limited recovery are insufficient to tackle global challenges. Awino and Apitz (2024) assert that implementing frameworks based on the waste hierarchy and CE is crucial for reversing unsustainable trends. These frameworks emphasize waste prevention, reduction, reuse, and recycling rather than disposal, facilitating resource efficiency and minimizing environmental damage. Incorporating CE ideas into waste systems offers a means to achieve sustainable practices that can diminish waste quantities, facilitate climate initiatives, and create new socioeconomic prospects. In this context, it is essential to address the solid waste concerns encountered by both developing and developed nations prior to the incorporation of CE concepts, as this offers an objective framework for resolving these issues. The following sections offer justification on solid waste challenges and the need to address it in developing economies.

1.3.1 Growing Volumes of Municipal Solid Waste

MSW is increasing due to urbanization, rising incomes, and changes in lifestyle. High-income nations produce greater waste per capita; yet, developing economies are experiencing the most accelerated growth, placing further pressure on their already vulnerable waste management systems. The unregulated proliferation of waste strain collection, transportation, and disposal capabilities, frequently resulting in open dumping and illicit incineration (World Bank, 2018). Global evaluations indicate that demographic and economic factors are elevating MSW levels, particularly in developing nations seeing the most significant increases (Chen et al., 2020). Analyses of the global context highlight that although per capita generation is greater in affluent economies, overall volumes in low- and middle-income nations are increasing most rapidly, intensifying strain on fragile infrastructure (Sharma & Jain, 2020). Furthermore, recent evaluations indicate that official estimates may underestimate the magnitude of the issue, as waste data in numerous countries is either incomplete or inconsistent, implying that worldwide MSW quantities could be considerably higher than stated (Maalouf & Mavropoulos, 2023).

1.3.2 Inequality in Waste Management Capacity

The capability for waste management is very disparate between regions. In developed countries, collection rates can exceed 90%, supported by advanced recycling and energy recovery systems. In contrast, many low- and middle-income countries see collection rates of approximately 50%, leading to significant amounts of waste staying uncollected or disposed of in open spaces (Wilson et al., 2006; World Bank, 2018). These differences are not merely technical but also rooted in broader inequality. Kalina (2020) asserts that discussions regarding waste and climate change must explicitly acknowledge the influence of inequality on vulnerability to waste-related hazards, with vulnerable populations disproportionately impacted by inadequate services. Kocak and Baglitas (2022) demonstrate that income inequality, alongside human development indices and energy efficiency, significantly influences the sustainability of MSW management systems. Insufficient financial resources, inadequate legal frameworks, and defective infrastructure exacerbate these inequities, forcing communities in the Global South to rely on open dumps that present environmental and public health hazards (Khatib, 2011; Abubakar et al., 2022). The prominence of informal participants in waste collection and recycling highlights the deficiencies of official institutions and the resilience of marginalized populations. Studies suggest that inclusive measures, such as the formation of waste picker cooperatives, may alleviate these gaps by improving livelihoods and promoting more sustainable CE pathways (Buch et al., 2021).

1.3.3 Environmental and Public Health Impacts

Insufficient SWM immediately leads to environmental deterioration and negative health effects. Open dumping pollutes soil and groundwater with heavy metals and other poisons, while uncontrolled combustion emits hazardous air pollutants that lead to respiratory and cardiovascular disorders (Manisalidis et al., 2020; Mitra et al., 2022). Unregulated landfills provide significant anthropogenic sources of methane, a powerful greenhouse gas, rendering SWM a crucial climate concern (UNEP, 2021). These environmental concerns result in considerable public health consequences, as evidenced by systematic reviews that associate inadequate MSW practices with increased rates of respiratory infections, diarrheal diseases, and vector-borne illnesses (Vinti et al., 2021). Public health hazards are particularly severe in informal settlements, where residents are near dumpsites and have disproportionate exposure to infections and hazardous pollutants (Chaturvedi & Khare, 2022). In addition to direct health effects, insufficient SWM exacerbates healthcare burdens and compromises sustainability, as health systems also contribute to environmental footprints (Lenzen et al., 2020). This information highlights the imperative of seeing waste management as both an environmental issue and a public health priority.

1.3.4 Emerging and Complex Waste Streams

The mix of global waste is changing swiftly, presenting new issues for waste management systems. In 2019, electronic waste (e-waste) totaled 53.6 million tons, with merely 17% being properly collected and recycled (Forti et al., 2020). In addition to its substantial quantity, e-waste comprises dangerous components, such as batteries that emit heavy metals and emerging pollutants, which can have detrimental environmental effects if mishandled (Melchor-Martínez et al., 2021). The discharge of plastic waste into rivers and oceans has initiated worldwide ecological catastrophes, jeopardizing biodiversity and food security (Jambeck et al., 2015). Pharmaceutical and medical wastes contribute intricate contaminants into aquatic systems, necessitating sophisticated biological treatment technology to reduce dangers (Khalidi-Idrissi et al., 2023). The accumulation of construction waste and hazardous industrial by-products persists, exacerbating the strain on vulnerable waste infrastructure. Simultaneously, these waste streams possess unexploited resource potential, especially in the recovery of essential metals for clean technologies, which recycling and CE methods can facilitate (Hagelüken & Goldmann, 2022). Collectively, these processes underscore the escalating complexity of waste, requiring comprehensive and novel strategies for sustainable management.

1.3.5 Socioeconomic Dimensions of Waste

In numerous developing economies, waste management is significantly reliant on the informal sector, whereby waste pickers are essential for recycling and material recovery. Although these occupations provide livelihoods, they frequently exhibit inadequate working conditions, insufficient recognition, and social vulnerability (Gutberlet & Uddin, 2017). Research indicates that waste production and recovery are significantly influenced by demographic and socioeconomic variables, with sustainability indices providing insights into the impact of inequality and development on waste streams (Van Fan et al., 2019). Open-access socioeconomic data indicate that income levels, poverty, and urban density are essential factors influencing waste patterns and system performance (Dunkel et al., 2022). Household income, lifestyle, and occupation similarly impact the volume and nature of waste generated, hence affecting the need for informal recycling labor (Kala & Bolia, 2020). Globally, machine learning analysis indicates that socioeconomic growth significantly influences SWM performance, with lower-income cities increasingly dependent on informal garbage workers to address service deficiencies (Velis et al., 2023). Incorporating these individuals into formal systems poses a difficulty yet also offers a chance to rectify imbalances and establish more inclusive waste value chains.

1.3.6 The Need for Sustainable Practices

The worldwide solid waste crisis is no longer a concealed problem limited to urban peripheries; it has emerged as one of the paramount sustainability problems of our era. The five authors previously identified global challenges: increasing waste volumes, disparities in waste management capacity, significant environmental and health repercussions, emerging complex waste streams, and socioeconomic vulnerabilities; all illustrate the insufficiency of linear, end-of-pipe waste management strategies. These traditional approaches, based on "collect–transport–dispose," have attained their limits. Landfills are inundated, open dumping and incineration persist, and communities in developing economies grapple with increasing expenses and heightened environmental hazards. This situation highlights the pressing necessity for sustainable practices based on the CE concept.

1.3.7 Responding to Growing Waste Volumes

Global MSW generation is anticipated to rise from 2.01 billion tons in 2016 to 3.40 billion tons by 2050 (World Bank, 2018). Population expansion, urban development, and increasing affluence are intensifying waste production, particularly in the Global South. Traditional landfill systems are unable to accommodate these trends, resulting in increasing environmental risks and cost burdens on municipal budgets. Sustainable techniques based on CE concepts provide answers by minimizing waste at the source and redirecting items from landfills. Recycling, composting, and waste-to-energy technologies prolong product lifecycles and generate value from waste materials. These techniques alleviate the strain on collection and disposal systems while fostering opportunities for resource recovery and the growth of local industries.

1.3.8 Reducing Inequalities in Waste Management Capacity

The international disparity in waste management capabilities illustrates profound structural imbalances. Developed nations generally attain collection rates exceeding 90%, bolstered by sophisticated recycling infrastructure, whereas most developing countries continue to grapple with rates around 50% (Wilson et al., 2006; World Bank, 2018). Inadequate financing, poor governance, and deficient infrastructure result in millions of tons of waste remaining uncollected, frequently discarded in rivers or along roadsides, or incinerated in the open air. These activities disproportionately impact on the most impoverished urban inhabitants, who reside in proximity to uncontrolled waste. Sustainable methods mitigate these inequities by offering scalable, decentralized solutions. Community-oriented composting initiatives, small-scale waste recovery operations, and cooperative recycling programs can enhance service accessibility in resource-limited environments. Furthermore, including circular practices into national and municipal policies enables governments to progressively shift from hazardous disposal methods while enhancing the resilience of local waste systems.

1.3.9 Mitigating Environmental and Public Health Risks

Insufficient SWM is both an environmental issue and a significant public health hazard. Open dumping pollutes soil and groundwater with heavy metals, whilst uncontrolled combustion emits harmful air pollutants that lead to respiratory ailments and cardiovascular diseases (Manisalidis et al., 2020; Mitra et al., 2022). Unregulated landfills provide significant anthropogenic sources of methane, a GHG that is 84 times more effective than carbon dioxide during a 20-year timeframe (UNEP, 2021). Public health hazards are especially pronounced in informal settlements, where exposure to waste-derived diseases and contaminants is most significant (Vinti et al., 2021). Sustainable practices provide healthier alternatives by minimizing uncontrolled dumping and burning. Diverting organic waste to composting or anaerobic digestion mitigates methane emissions while generating soil amendments and biogas. Implementing secure disposal solutions for hazardous and medical waste mitigates harmful exposure and diminishes disease loads. Aligning waste management with environmental health concerns ensures that sustainable methods safeguard both ecosystems and human populations.

1.3.10 Addressing Emerging and Complex Waste Streams

The composition of global waste is changing in ways that surpass standard management strategies. In 2019, e-waste amounted to 53.6 million tons, with merely 17% undergoing formal recycling (Forti et al., 2020). Plastics persist in contaminating rivers and oceans, resulting in ecological catastrophes that jeopardize biodiversity and food systems (Jambeck et al., 2015). The increase in construction debris, toxic waste, and pharmaceutical waste is further burdening the already vulnerable waste infrastructure. CE-based methodologies are distinctly equipped to tackle this complexity by advocating for product redesign, EPR, and eco-innovation. Eco-design guidelines can diminish the toxicity of electronics, but EPR schemes mandate companies to manage the collection and recycling of plastics, batteries, and packaging. Circular solutions mitigate the risks associated with complex waste while also harnessing the resource possibilities contained inside, including vital metals from electronics and reusable construction materials.

1.3.11 Tackling Socioeconomic Vulnerabilities

In numerous emerging economies, waste management systems are predominantly dependent on the informal sector. Waste pickers are essential to recycling and material recovery; nonetheless, their labor is characterized by unstable conditions, insufficient social protection, and minimal acknowledgment (Gutberlet & Uddin, 2017). Notwithstanding their marginalization, informal players frequently serve as the most effective recyclers, redirecting substantial quantities of waste from landfills. Sustainable practices prioritize inclusion and formalization, converting socioeconomic vulnerabilities into possibilities for empowerment. Cooperative models in Brazil and India demonstrate that integrating waste pickers into groups strengthens negotiating power, secures equitable remuneration, and improves system efficiency (Buch et al., 2021). Integrating informal laborers into formal waste governance frameworks enhances livelihoods while simultaneously bolstering recycling rates and material recovery in CE activities. This renders waste management both socially equitable and environmentally sustainable.

1.3.12 Aligning with Global Sustainability Goals

Sustainable waste techniques ultimately harmonize with and bolster global development priorities. By diminishing volumes via recycling and recovery, they immediately contribute to SDG 12 (Responsible Consumption and Production). By establishing inclusive green employment and diminishing disparities in service delivery, they promote SDG 8 (Decent Work and Economic Growth) and SDG 10

(Reduced Inequalities). By tackling urban waste challenges and enhancing service delivery, they contribute to Sustainable Development Goal 11 (sustainable cities and communities). By mitigating methane and other greenhouse gas emissions, they are crucial in achieving Sustainable Development Goal 13 (Climate Action). Circular waste management is both a technological imperative and a strategic facilitator of sustainable development.

1.4 DEVELOPING ECONOMIES

1.4.1 Facets of Developing Economies

Developing economies are characterized by various interconnected structural, social, institutional, and environmental elements that influence their growth paths and integration into the global economy. In contrast to industrialized economies, which are generally defined by diversified industries, elevated productivity, and established institutions, developing economies encounter distinct difficulties and opportunities. Their economic and social realities must be comprehended collectively, as various factors interact dynamically to shape trajectories towards sustainable development. Gonos et al. contend that aspects of growth, including economic structure, governance, and social systems, are intricately linked, with advancements in one domain frequently reliant on enhancements in others.

The economic structure is a crucial aspect of developing economies. A significant number continues to rely extensively on primary sectors, including agriculture, mining, and resource extraction. Although these sectors generate employment and foreign exchange revenue, excessive dependence on commodities renders nations vulnerable to fluctuations in global markets. Zambia obtains over 70% of its export revenues from copper, rendering the nation susceptible to variations in global copper prices (World Bank, 2022). Inadequate industrialization and feeble manufacturing sectors constrain potential for value enhancement and diversification, leading to enduring structural imbalances (Gelb et al., 2014).

A secondary aspect is the demographic characteristics of developing economies. These nations frequently possess youthful and swiftly expanding populations, which can yield a demographic dividend if sufficiently bolstered by investments in education and skill development. Nonetheless, population expansion exerts strain on social services, infrastructure, and labor markets. Accelerated urbanization exacerbates this complexity, as migration to metropolitan areas frequently surpasses the advancement of housing, transportation, and sanitary infrastructures. The outcome is the proliferation of informal settlements, affecting public health and urban planning (Fox, 2014).

The institutional and governance framework constitutes another essential aspect. Numerous developing economies have obstacles including fragile institutions, policy volatility, and corruption, which hinder efficient resource management and deter investment. Simultaneously, international assistance and donor impact can dictate development priorities, occasionally fostering dependency instead of autonomy in policymaking (Kaufmann et al., 2010). Robust governance frameworks are crucial to realize the promise of various aspects, including industrial growth, financial inclusion, and social development.

The aspect of infrastructure and technology is closely intertwined. In numerous developing economies, infrastructural deficiencies such as insufficient roads, inconsistent energy, and restricted access to clean water impede production and competitiveness (Calderón & Servén, 2010). However, these nations exhibit considerable potential for technological leapfrogging. Mobile banking platforms like M-Pesa in Kenya demonstrate how innovation can circumvent conventional systems to enhance financial inclusion (Donovan, 2012). The interaction between infrastructural shortcomings and technological possibilities demonstrates the limitations and innovative potential of these economies.

Additional essential aspects encompass commerce and global integration, financial systems, social growth, and environmental sustainability. Reliance on a limited array of exports, inadequate financial inclusion, ongoing poverty and inequality, and significant susceptibility to climate change are prevalent issues. Simultaneously, informality persists as a predominant characteristic of economic existence, with Barbosa et al. (2021) demonstrating how socioeconomic factors and mobility patterns compel substantial segments of the population into informal employment. These actions are essential for survival and resilience in environments of structural limitation.

The aspects of developing economies include economic, demographic, institutional, infrastructural, social, financial, and environmental factors. The factors are interdependent and mutually reinforcing, rendering development a complicated and multidimensional process. Comprehending these aspects is essential for formulating policies that foster inclusive, resilient, and sustainable development in the Global South.

1.4.2 Economic Structure

The economic framework of developing nations is frequently characterized by fundamental sectors, including agriculture, mining, and resource exploitation. These sectors generate employment and foreign exchange but also introduce vulnerability to swings in commodity prices. Zambia generates more than 70% of its export revenues from copper, rendering its economy highly susceptible to fluctuations in global copper prices (World Bank, 2022). Inadequate industrialization and feeble manufacturing capabilities obstruct value enhancement and economic diversification, perpetuating structural imbalances (Gelb et al., 2014). Gonos et al. (2022) assert that dependence on natural resources without industrial enhancement limits resilience and sustainable growth. Erokhin and Gao (2020) demonstrate that the COVID-19 pandemic affected trade and food security in 45 developing nations, particularly affecting those reliant on agricultural production and food imports. Their findings emphasize that insufficient industrial diversification renders countries incapable of absorbing external shocks, hence accentuating the vulnerability of development models reliant on primary sector predominance.

1.4.3 Demographics

Emerging economies generally possess youthful and swiftly expanding populations, a distinctive demographic trait that offers both prospects and obstacles. Africa is anticipated to double its population by 2050, with over 60% of individuals under the age of 25 (UN DESA, 2019). This demographic profile presents the possibility for a demographic dividend, wherein a substantial working-age population can stimulate innovation, enhance productivity, and facilitate economic transformation. The actualization of this potential relies on intentional investments in education, healthcare, and employment generation. In the absence of such investments, population expansion may intensify unemployment, poverty, and inequality. Kozlovskyi et al. (2020) present empirical evidence about the dual influence of demographic determinants on economic results. Their research indicates that although a youthful labor force can contribute to sustained economic growth, in developing nations it frequently becomes a liability when fragile institutions, poor labor markets, and insufficient infrastructure are unable to accommodate the burgeoning workforce. This directly supports the fact that urban migration often exceeds the growth of housing, transportation, and public facilities, resulting in the emergence of informal settlements and social tensions (Fox, 2014).

Roy (2021) asserts that demographics affect practically all facets of economics, finance, and policy. He observes that demographic age composition, birth rates, and migration trends influence both innovation and fiscal viability. In Africa, a youthful population may stimulate entrepreneurship and technology adoption; yet elevated dependency ratios and restricted job prospects provide a danger of instability if not accompanied by inclusive economic policies. Roy's argument emphasizes the significance of investing in human capital, contending that nations that neglect to educate and employ its youth jeopardize transforming a potential advantage into a demographic burden.

In summary, demography in developing economies serves as both a catalyst for opportunity and a source of fragility. Leveraging this potential necessitates proactive policies that enhance educational access, fortify health systems, and provide sustainable employment, so converting population expansion into a true catalyst for development.

1.4.4 Institutions and Governance

Institutional quality and governance are essential factors influencing economic performance in developing economies. Inadequate regulatory frameworks, corruption, and policy volatility deter investment and hinder developmental results, whereas robust institutions can enhance efficiency and attract resources that promote growth. Kaufmann and Kraay (2016) contend that governance deficiencies diminish the efficient utilization of public resources and constrain the efficacy of development methods. This issue is especially pronounced in numerous emerging nations, where donor-driven agendas frequently dictate national goals, fostering cycles of dependency instead of encouraging self-sufficiency. Nonetheless, there are success narratives, shown by Rwanda and Botswana, illustrating that governance reforms centered on transparency, accountability, and institutional fortification can bolster investor confidence and promote equitable growth. Research further demonstrates the significance of governance in influencing sectoral outcomes. Tan and Taeihagh (2020) demonstrate in their assessment of smart city administration in developing nations that fragmented and feeble institutional frameworks impede governmental capacity to oversee modernization programs. In the absence of cohesive and accountable governance, even well-financed infrastructure or technology initiatives fail to provide inclusive advantages. Nguyen and Su (2022) emphasize the impact of governance quality on the efficacy of public expenditure. Their research on energy poverty in emerging nations revealed that although government expenditure can mitigate poverty, its effectiveness is markedly diminished in environments characterized by inefficient institutions and inadequate governance. This discovery confirms the assertion that governance deficiencies hinder states' capacity to tackle urgent social and developmental issues.

The environmental aspect of governance is equally vital. Wawrzyniak and Doryń (2020) illustrate that institutional quality influences the correlation between economic growth and environmental sustainability. Robust institutions facilitate progress without increasing carbon emissions, whereas inadequate governance intensifies the conflict between development and environmental damage. Collectively, these studies underscore that strong governance serves as both a catalyst for investment and economic stability, and a protector of social inclusiveness and environmental sustainability.

1.4.5 Infrastructure and Technology

Infrastructure deficiencies persist as a primary limitation in numerous developing economies, where inadequate road systems, inconsistent electrical supply, and restricted access to clean water elevate corporate costs and diminish competitiveness. Calderón and Servén (2010) illustrate that infrastructure investment is significantly associated with long-term growth and poverty alleviation, highlighting the critical role of roads, energy, and water systems as facilitators of economic transformation. Numerous emerging nations still have a substantial infrastructure deficit, which hinders industrialization and restricts engagement in global value chains.

Simultaneously, technical breakthroughs are facilitating leapfrogging, overcoming conventional development phases via digital transformation. Mobile banking services, exemplified by Kenya's M-Pesa, have transformed financial inclusion, enabling millions without access to traditional banks to participate in transactions and savings (Donovan, 2012). Oloyede et al. (2023) demonstrate that the digital economy significantly enhances growth in developing contexts by facilitating e-commerce, digital finance, and knowledge transfer, despite the lack of substantial physical infrastructure. Demaidi (2025) emphasizes that the implementation of artificial intelligence (AI) initiatives can enhance service delivery, improve healthcare access, and increase industrial efficiency, hence assisting developing nations in closing structural disparities.

Nonetheless, the advantages of technology are not inherent. Tan and Taeihagh (2020) warn that inadequate governance frameworks frequently diminish the efficacy of digital and "smart city" initiatives in developing economies. In the absence of robust coordination, accountability, and inclusive governance, digital solutions may exacerbate disparities instead of alleviating them. Similarly, Hadian et al. (2021) demonstrate in their examination of medical tourism that both infrastructure and technology are essential for competitiveness in global service sectors, since insufficient transportation, health, and ICT infrastructure limits potential despite increasing innovation.

Collectively, these lessons emphasize that although infrastructure is essential for development, technology offers innovative solutions to surmount obstacles. With strong governance and inclusive initiatives, digital breakthroughs may expedite growth, enhance competitiveness, and bolster resilience in developing countries.

1.4.6 Social and Environmental Dimensions

Social growth in numerous developing economies continues to be inconsistent, characterized by enduring poverty, inequality, and restricted access to healthcare and education. The World Bank (2020) predicts that sub-Saharan Africa comprises approximately 60% of the global extreme poor, indicative of structural impediments that hinder equitable progress. Gender differences exacerbate these issues, as women frequently encounter barriers to entrepreneurship, formal employment, and education (Chant & Sweetman, 2012). Such disparities not only impede individual welfare but also diminish national production and resilience. Rai et al. (2021) emphasize that crises like the COVID-19 pandemic exacerbate these vulnerabilities, illustrating that the absence of robust social safety nets and inclusive systems can swiftly destabilize social and economic equilibrium.

Environmental stresses further intensify these susceptibilities. Numerous resource-dependent economies encounter the "resource curse," wherein natural riches engender dependence instead of diversification, concurrently confronting significant vulnerability to climate change effects, including droughts, floods, and food insecurity (Auty, 2001). Hariram et al. (2023) present a comprehensive framework of "sustainalism" that underscores the interdependence of social, economic, and environmental sustainability. Their viewpoint endorses the notion that addressing poverty, inequality, and climate change necessitates comprehensive and coordinated strategies that acknowledge these interrelations.

A crucial aspect is informality, which continues to be widespread in developing economies. Barbosa et al. (2021) demonstrate how socioeconomic factors and patterns of human mobility compel millions to engage in informal employment, characterized by precarious and unregulated livelihoods. Simultaneously, environmental issues like inadequate waste management are linked to insufficient institutional capacity and a lack of public awareness. Debrah et al. (2021) emphasize that fostering sustainability via formal education and awareness initiatives is crucial in developing contexts, as it can enable communities to adopt ecologically responsible activities. In summary, the social and environmental aspects of developing economies reveal significant weaknesses as well as opportunities. Confronting these challenges necessitates cohesive measures that enhance resilience, foster equity, and connect environmental sustainability with inclusive social and economic policies.

1.5 SWM IN DEVELOPING ECONOMIES

SWM in developing economies continues to be a significant concern due to accelerated urbanization, population expansion, and inadequate institutional frameworks.

In contrast to wealthy nations, which possess established systems for collection, recycling, and disposal, poor countries frequently depend on fragmented, underfunded, and informal arrangements (Azevedo et al., 2021). The World Bank (2018) forecasts that worldwide waste production will increase to 3.4 billion tons by 2050, with low-income nations exhibiting the most significant growth rates. In sub-Saharan Africa and South Asia, collection efficiency frequently remains under 50%, resulting in substantial quantities of uncollected waste (Kaza et al., 2018). A characteristic attribute of waste in developing nations is its substantial organic component, typically ranging from 50% to 70%, rendering it appropriate for composting and biogas production (Guerrero et al., 2013). Nonetheless, inadequate infrastructure and insufficient source segregation diminish recovery rates, despite the significant potential of technology for energy recovery from organic materials (Khan et al., 2022). Marginalized informal waste pickers are essential in recycling and alleviating municipal burdens (Wilson et al., 2012). Municipalities encounter significant budgetary limitations, with SWM generally accounting for 20–50% of local budgets, yet still inadequately providing service delivery (Hoornweg & Bhada-Tata, 2012). The repercussions are extensive: open dumpsites, unchecked incineration, and inadequate disposal methods pollute soil, air, and water, hence exposing communities to vector-borne diseases (Nzediegwu & Chang, 2020). Enhancing awareness by education and community involvement can promote source waste separation and reinforce sustainable behaviors (Debrah et al., 2021). Opportunities remain to transition from a linear "collect–dump" approach to CE practices that prioritize reduction, reuse, and recycling (Kirchherr et al., 2017). Enhancing legal frameworks, investing in infrastructure, incorporating informal workers, and fostering behavioral change are vital for lasting solutions. Consequently, SWM in developing economies constitutes both a significant barrier and an opportunity to promote inclusive green growth. Strategic initiatives can convert waste from an environmental burden into a resource that enhances livelihoods, promotes health, and fosters. Thus, solid waste management (SWM) in emerging economies both a substantial developmental challenge and a considerable opportunity to promote inclusive green growth. With the backing of suitable policies, investment, and institutional capacity, strategic solid waste management interventions can convert waste from an environmental burden into a valuable resource that creates employment, enhances public health, conserves natural resources, diminishes greenhouse gas emissions, and promotes circular economy practices. These methods foster sustainable livelihoods, improve resource efficiency, and advance resilient and inclusive socio-economic development.

1.5.1 High Waste Generation with Limited Collection

The rapid urbanization and population increase in developing economies have markedly intensified the output of MSW, while collection efficiency continues to fall short of demand. In numerous cities across Africa and Asia, less than 50% of waste materials are collected, resulting in the accumulation of uncollected waste in open areas, streets, or informal dumps (Kaza et al., 2018). This dilemma is exacerbated by evolving consumption patterns that augment nonbiodegradable waste sources (Sharma & Jain, 2020). Insufficient municipal resources, poor truck availability, and deficient planning frameworks further restrict collection capability. Evidence from Johannesburg indicates that institutional deficiencies and the absence of integrated governance frameworks intensify inefficiencies in SWM, highlighting the necessity for institutional fortification and policy reforms (Kubanza & Simatele, 2020). Recent global evaluations indicate that waste generation is increasing more rapidly than previously anticipated, particularly in low- and middle-income nations, underscoring the necessity for innovative and community-oriented solutions to address the collection deficit (Maalouf & Mavropoulos, 2023).

1.5.2 Dominance of Organic Waste

In developing economies, MSW predominantly consists of organic material, usually comprising 50–70%, derived from food and agricultural by-products (Guerrero et al., 2013). This composition facilitates resource recovery by composting and anaerobic digestion; nevertheless, inadequate source segregation frequently leads to the contamination of organics with plastics and hazardous waste, hence diminishing their value (World Bank, 2018). Data from Bangladesh indicates that anaerobic digestion of organic waste can produce biogas and hydrogen, offering renewable energy alternatives and decreasing dependence on fossil fuels (Khan, 2020). Research in India indicates that inadequate infrastructure, insufficient awareness, and ineffective waste segregation impede efficient food waste management, notwithstanding its potential for compost and bioenergy generation (Sahoo et al., 2024). Mitigating these obstacles by localized composting systems, biogas facilities, and focused awareness initiatives could convert organic waste into a significant resource for sustainable agriculture and renewable energy production.

1.5.3 Informal Sector Involvement

The informal sector is a crucial contributor to waste recovery in developing economies, with waste pickers and cooperatives accumulating approximately 20–30% of recyclable materials in certain urban regions (Wilson et al., 2012). Despite these contributions, informal laborers generally operate under hazardous conditions, lacking protective equipment or social security, and continue to be excluded from municipal planning (Harfadli et al., 2024). Their marginalization compromises both social justice and environmental sustainability. Evidence from Johannesburg indicates that institutional deficiencies frequently hinder municipalities from establishing formal engagement with the informal sector, even though these actors alleviate landfill congestion and decrease municipal expenses (Kubanza & Simatele, 2020). Integrating waste pickers into formal waste management frameworks via cooperatives, specialized training, and provision of protective equipment can improve recycling rates, generate dignified green employment opportunities, and reinforce sustainable waste governance. Recognition of their function is thus not solely a matter of efficiency but also a matter of social justice and inclusivity.

1.5.4 Limited Financial Resources

SWM represents a significant financial burden for municipalities, frequently accounting for 20–50% of urban budgets in low-income nations. Despite this expenditure, many cities struggle to provide universal collection services or to invest in sanitary landfills and recovery facilities (Hoornweg & Bhada-Tata, 2012). Financing constraints stem from restricted tax bases, inadequate revenue collection, and competing priorities, including health and education. Research conducted in Johannesburg indicates that these challenges are exacerbated by insufficient institutional frameworks, where poor governance restricts the capacity to effectively mobilize and manage resources (Kubanza & Simatele, 2020). Innovative approaches such as extended producer responsibility (EPR), user fees, and donor-supported infrastructure are essential for addressing funding gaps (Moqsud et al., 2011). Insights from Germany illustrate the enhancement of financial sustainability through the adaptation of international best practices to specific local contexts (Azevedo et al., 2021). Furthermore, awareness campaigns and educational initiatives decrease waste volumes, thereby indirectly alleviating financial pressures on municipalities (Debrah et al., 2021).

1.5.5 Environmental and Health Risks

Insufficient SWM in developing economies subjects communities to various environmental and health risks. Open dumping and uncontrolled burning emit toxic pollutants and greenhouse gases, whereas leachate contaminates both surface and groundwater (Moqsud et al., 2011). Waste pickers, as vulnerable groups, experience a disproportionate impact, encountering elevated risks of respiratory infections, diarrhea, and vector-borne diseases (Abubakar et al., 2022; Wilson et al., 2012). The improper disposal of hazardous materials, such as electronic waste and medical waste, intensifies environmental degradation and poses significant public health risks (Maniruzzaman et al., 2022). The identified risks indicate that inadequate waste management constitutes both an environmental issue and a critical public health concern. Addressing these issues necessitates the rigorous enforcement of environmental regulations, investment in sanitary landfills and safe treatment facilities, and the implementation of targeted health protection measures for at-risk populations. Sustainable waste governance effectively mitigates ecological damage and reduces disease burdens.

1.6 SOCIOECONOMIC AND ENVIRONMENTAL EFFECTS OF MEDIOCRE WASTE MANAGEMENT

SWM is a crucial factor in sustainable urban development; however, in numerous developing economies, management practices are often inadequate, inconsistent, and fragmented. Mediocre waste management, marked by inadequate collection efficiency, insufficient segregation, and dependence on open dumping, results in significant socioeconomic and environmental repercussions. Research indicates that inadequate waste governance intensifies public health hazards and environmental deterioration while also compromising social welfare and economic development (Jagun et al., 2023).

Inefficient waste systems result in significant financial burdens for municipalities, yielding minimal returns in service delivery from a socioeconomic standpoint. Numerous cities dedicate substantial budgetary resources to SWM; however, inefficiencies lead to uncollected waste and informal dumping practices. This results in a detrimental cycle in which heightened waste volumes strain inadequately funded systems, thereby diminishing overall productivity (Velis et al., 2023). Research indicates that socioeconomic variables, including income levels, education, and consumption patterns, directly affect the type and quantity of waste generated, resulting in disparities in management effectiveness (Kala & Bolia, 2020). Higher-income communities generate greater amounts of packaging and electronic waste, whereas lower-income communities produce larger proportions of organic waste, necessitating distinct management strategies. In the absence of targeted policies, these disparities intensify inequalities in service provision.

Furthermore, inadequate waste management sustains reliance on the informal sector. Informal waste pickers play a crucial role in recycling and material recovery; however, their exclusion from formal systems leads to precarious livelihoods, adverse health outcomes, and insufficient social protection (Jagun et al., 2023). Mediocre SWM reinforces socioeconomic vulnerabilities instead of facilitating opportunities for green employment and inclusive growth.

The environmental impacts are similarly significant. The practice of open dumping and uncontrolled burning results in the release of greenhouse gases and toxic emissions, thereby exacerbating climate change and air pollution. Leachate from unmanaged sites contaminates groundwater, posing risks to biodiversity and human health (Turyahabwe et al., 2022). Case studies from Uganda indicate that inadequate management of municipal waste adversely affects water bodies and agricultural lands, thereby reducing environmental resilience. In the construction sector, inadequate waste management leads to significant debris accumulation, resulting in pollution of land and water systems, increased project costs, and diminished efficiency (Tafesse et al., 2022). This demonstrates that inadequate waste governance transcends municipal systems, impacting various sectors of the economy and resulting in cascading socio-environmental consequences.

Machine learning analyses at the global level indicate that socioeconomic development significantly influences SWM performance, suggesting that nations with weaker economic frameworks are more likely to exhibit mediocre waste management practices (Velis et al., 2023). This indicates that waste challenges are fundamentally rooted in structural inequalities, necessitating extensive reforms. Addressing these issues requires investment in infrastructure, integration of informal workers, public education, and adoption of CE strategies.

Mediocre waste management exacerbates inequality, undermines economic productivity, and accelerates environmental degradation. The evidence highlights that effective SWM is not only a technical requirement but also a socioeconomic and environmental necessity. Transforming waste systems into inclusive, sustainable, and resource-efficient models is essential for enhancing resilience, alleviating poverty, and achieving sustainable urban development in developing economies.

SWM in developing economies constitutes a significant socioeconomic and environmental challenge. Rapid urbanization, population growth, and inadequate institutional frameworks have led to significant waste generation, with collection efficiency frequently falling below 50% in cities across Africa and Asia. Organic waste constitutes 50–70% of municipal streams; however, inadequate segregation diminishes the potential for composting and biogas recovery. Informal waste pickers play a crucial role in recycling and landfill diversion; however, their exclusion from formal systems renders them vulnerable and undervalued. Municipalities encounter significant financial limitations, as SWM accounts for up to 50% of city budgets, yet universal coverage remains unachieved due to restricted tax bases and competing priorities.

The implications of mediocre waste management reach beyond mere inefficiencies. Inadequately managed waste leads to GHG emissions, groundwater contamination, and significant health risks, including respiratory infections, diarrhea, and vector-borne diseases, especially affecting children and informal workers. Socioeconomic disparities influence waste composition and management, and insufficient governance intensifies inequality and environmental degradation. Evidence indicates that the integration of informal workers, the adoption of innovative financing models, investment in infrastructure, and the enhancement of public awareness can significantly transform SWM systems. Reforms grounded in CE principles can transform waste into a resource, create green jobs, enhance health outcomes, and foster inclusive sustainable development.

REFERENCES

Abubakar, I.R., Maniruzzaman, K.M., Dano, U.L., AlShihri, F.S., AlShammari, M.S., Ahmed, S.M.S., Al-Gehlani, W.A.G. and Alrawaf, T.I. (2022) 'Environmental sustainability impacts of solid waste management practices in the global South', *International Journal of Environmental Research and Public Health*, 19(19), p. 12717.

Abubakar, I.R., Maniruzzaman, K.M., Dano, U.L., AlShihri, F.S., AlShammari, M.S., Ahmed, S.M.S., Al-Gehlani, W.A.G. and Alrawaf, T.I. (2022) 'Environmental sustainability impacts of solid waste management practices in the Global South', *International Journal of Environmental Research and Public Health*, 19(19), Article 12717. https://doi.org/10.3390/ijerph191912717.

Armendáriz, B. and Morduch, J. (2010) *The Economics of Microfinance*. 2nd edn. Cambridge, MA: MIT Press.

Auty, R.M. (2001) *Resource Abundance and Economic Development*. Oxford: Oxford University Press.

Awan, U., Kanwal, N. and Bhutta, M.K.S. (2020) 'A literature analysis of definitions for a circular economy', in M.K.S. Bhutta, U. Awan and S. Ren (eds.), *Logistics Operations and Management for Recycling and Reuse*. Cham: Springer, pp. 19–34. https://doi.org/10.1007/978-3-030-40831-0_2.

Awino, F.B. and Apitz, S.E. (2024) 'Solid waste management in the context of the waste hierarchy and circular economy frameworks: An international critical review', *Integrated Environmental Assessment and Management*, 20(1), pp. 9–35.

Azevedo, B.D., Scavarda, L.F., Caiado, R.G.G. and Fuss, M. (2019) 'Improving urban household solid waste management in developing countries based on the circular economy: A case study in Brazil', *Journal of Cleaner Production*, 231, pp. 189–199.

Azevedo, B.D., Scavarda, L.F., Caiado, R.G.G. and Fuss, M. (2021) 'Improving urban household solid waste management in developing countries based on the German experience', *Waste Management*, 120, pp. 772–783.

Barbosa, H., Hazarie, S., Dickinson, B., Bassolas, A., Frank, A., Kautz, H., Sadilek, A., Ramasco, J.J. and Ghoshal, G. (2021) 'Uncovering the socioeconomic facets of human mobility', *Scientific Reports*, 11(1), p. 8616.

Bellezoni, R.A., Adeogun, A.P., Paes, M.X. and de Oliveira, J.A.P. (2022) 'Tackling climate change through circular economy in cities', *Journal of Cleaner Production*, 381, p. 135126.

Buch, R., Marseille, A., Williams, M., Aggarwal, R. and Sharma, A. (2021) 'From waste pickers to producers: An inclusive circular economy solution through development of cooperatives', *Sustainability*, 13(16), p. 8925.

Calderón, C. and Servén, L. (2010) 'Infrastructure and economic development in Sub-Saharan Africa', *Journal of African Economies*, 19(suppl_1), pp. i13–i87.

Chant, S. and Sweetman, C. (2012) 'Fixing women or fixing the world? 'Smart economics', efficiency approaches, and gender equality in development', *Gender & Development*, 20(3), pp. 517–529. https://doi.org/10.1080/13552074.2012.731812.

Chaturvedi, S. and Khare, A. (2022) 'Solid wastes: Characteristics, composition and adverse effects on environment and public health', *Asian Journal of Advanced Research and Reports*, 16(7), pp. 9–30.

Chen, D.M.C., Bodirsky, B.L., Krueger, T., Mishra, A. and Popp, A. (2020) 'The world's growing municipal solid waste: Trends and impacts', *Environmental Research Letters*, 15(7), p. 074021.

Debrah, J.K., Vidal, D.G. and Dinis, M.A.P. (2021) 'Raising awareness on solid waste management through formal education for sustainability', *Recycling*, 6(1), p. 6.

Demaidi, M.N. (2025) 'Artificial intelligence national strategy in a developing country', *AI & Society*, 40(2), pp. 423–435.

Donovan, K. (2012) 'Mobile money for financial inclusion', in T. Kelly and C. Rossotto (eds.), *Information and Communications for Development 2012: Maximizing Mobile*. Washington, DC: World Bank, pp. 61–73.

Dumlao-Tan, M.I. and Halog, A. (2017) 'Moving towards a circular economy in solid waste management', in Ghosh, S.K. (ed.) *Advances in Solid and Hazardous Waste Management*. Cham: Springer, pp. 29–48.

Dunkel, J., de León, A.A., Călin, A.C. and Ciupe, A. (2022) 'Solid waste analysis using open-access socio-economic data for municipal waste management', *Sustainability*, 14(3), p. 1233. https://doi.org/10.3390/su14031233.

Ellen MacArthur Foundation (2013) *Towards the Circular Economy: Economic and Business Rationale for an Accelerated Transition*. Vol. 1. Cowes: Ellen MacArthur Foundation.

Elroi, H., Zbigniew, G., Agnieszka, W.C. and Piotr, S. (2023) 'Enhancing waste resource efficiency: Circular economy for sustainability and energy conversion', *Frontiers in Environmental Science*, 11, p. 1303792.

Erokhin, V. and Gao, T. (2020) 'Impacts of COVID-19 on trade and economic aspects of food security: Evidence from 45 developing countries', *International Journal of Environmental Research and Public Health*, 17(16), p. 5775. https://doi.org/10.3390/ijerph17165775.

Figge, F., Thorpe, A.S. and Gutberlet, M. (2023) 'Definitions of the circular economy: Circularity matters', *Ecological Economics*, 208, p. 107823.

Forti, V., Baldé, C.P., Kuehr, R. and Bel, G. (2020) *The Global E-waste Monitor 2020*. Bonn/Geneva/Rotterdam: United Nations University, ITU & ISWA.

Fox, S. (2014) 'The political economy of slums: Theory and evidence from Sub-Saharan Africa', *World Development*, 54, pp. 191–203. https://doi.org/10.1016/j.worlddev.2013.08.005.

Gallego-Schmid, A., Chen, H.M., Sharmina, M. and Mendoza, J.M.F. (2020) 'Links between circular economy and climate change mitigation', *Journal of Cleaner Production*, 260, p. 121115.

Geisendorf, S. and Pietrulla, F. (2018) 'The circular economy and circular economic concepts', *Thunderbird International Business Review*, 60(5), pp. 771–782.

Gelb, A., Meyer, C. and Ramachandran, V. (2014) 'Development as diffusion: Manufacturing productivity and Sub-Saharan Africa's missing middle', *Center for Global Development Working Paper* No. 357, Washington, DC: Center for Global Development.

Guerrero, L.A., Maas, G. and Hogland, W. (2013) 'Solid waste management challenges for cities in developing countries', *Waste Management*, 33(1), pp. 220–232.

Gutberlet, J. and Uddin, S.M.N. (2017) 'Household waste and health risks affecting waste pickers', *International Journal of Occupational and Environmental Health*, 23(4), pp. 299–310.

Hadian, M., Jabbari, A., Mousavi, S.H. and Sheikhbardsiri, H. (2021) 'Medical tourism development: A systematic review of economic aspects', *International Journal of Healthcare Management*, 14(2), pp. 576–582. https://doi.org/10.1080/20479700.2019.1677977.

Hagelüken, C. and Goldmann, D. (2022) 'Recycling and circular economy', *Mineral Economics*, 35(3), pp. 539–562.

Hailemariam, A. and Erdiaw-Kwasie, M.O., 2023. Towards a circular economy: Implications for emission reduction and environmental sustainability. *Business Strategy and the Environment, 32*(4), 1951–1965.

Harfadli, M.M., Ramadan, B.S., Rachman, I. and Matsumoto, T. (2024) 'Challenges and characteristics of the informal waste sector in developing countries: An overview', *Journal of Material Cycles and Waste Management*. https://doi.org/10.1007/s10163-024-01929-3.

Hariram, N.P., Mekha, K.B., Suganthan, V. and Sudhakar, K. (2023) 'Sustainalism: An integrated socio-economic-environmental model to address sustainable development and sustainability', *Sustainability*, 15(13), p. 10682. https://doi.org/10.3390/su151310682.

Hoornweg, D. and Bhada-Tata, P. (2012) *What a Waste: A Global Review of Solid Waste Management*. Washington, DC: World Bank.

Jagun, Z.T., Daud, D., Ajayi, O.M., Samsudin, S., Jubril, A.J. and Abdul Rahman, M.S. (2023) 'Waste management practices in developing countries: A socio-economic perspective', *Environmental Science and Pollution Research*, 30, pp. 116644–116655. https://doi.org/10.1007/s11356-022-21990-5.

Jambeck, J.R., Geyer, R., Wilcox, C., Siegler, T.R., Perryman, M., Andrady, A., Narayan, R. and Law, K.L. (2015) 'Plastic waste inputs from land into the ocean', *Science*, 347(6223), pp. 768–771.

Kabir, Z. and Kabir, M. (2021) 'Solid waste management in developing countries: Towards a circular economy', In: Baskar, C., Ramakrishna, S., Baskar, S., Sharma, R., Chinnappan, A., Sehrawat, R. (Eds.), *Handbook of Solid Waste Management: Sustainability through Circular Economy*. Singapore: Springer Singapore, pp. 1–34.

Kala, K., Bolia, N.B. and Sushil (2020) 'Waste management communication policy for effective citizen awareness', *Journal of Policy Modeling*, 42(3), pp. 661–678. https://doi.org/10.1016/j.jpolmod.2020.01.012.

Kalina, M. (2020) 'Against waste: A critique of the circular economy', *Ephemera: Theory and Politics in Organization*, 20(1), pp. 109–131.

Kaufmann, D., Kraay, A. and Mastruzzi, M. (2010) *The Worldwide Governance Indicators: Methodology and Analytical Issues*. World Bank Policy Research Working Paper No. 5430. Washington, DC: World Bank.

Kaza, S., Yao, L., Bhada-Tata, P. and Van Woerden, F. (2018) *What a Waste 2.0: A Global Snapshot of Solid Waste Management to 2050*. Washington, DC: World Bank. https://doi.org/10.1596/978-1-4648-1329-0.

Khalidi-Idrissi, A., Madinzi, A., Anouzla, A., Pala, A., Mouhir, L., Kadmi, Y. and Souabi, S. (2023) 'Recent advances in the biological treatment of wastewater rich in emerging pollutants produced by pharmaceutical industrial discharges', *International Journal of Environmental Science and Technology*, 20, pp. 1–22.

Khan, I. (2020) 'Waste to biogas through anaerobic digestion: Hydrogen production potential in the developing world – A case of Bangladesh', *International Journal of Hydrogen Energy*, 45(31), pp. 15951–15962. https://doi.org/10.1016/j.ijhydene.2020.04.112.

Khan, M.S., Mubeen, I., Caimeng, Y., Al-Ghamdi, S.G. and Khan, E. (2022) 'Waste-to-energy incineration technology: Recent development under climate change scenarios', *Waste Management & Research*, 40(12), pp. 1708–1729. https://doi.org/10.1177/0734242X221105312.

Kirchherr, J., Reike, D. and Hekkert, M. (2017) 'Conceptualizing the circular economy', *Resources, Conservation and Recycling*, 127, pp. 221–232.

Kirchherr, J., Yang, N.H.N., Schulze-Spüntrup, F., Heerink, M.J. and Hartley, K. (2023) 'Conceptualizing the circular economy revisited', *Resources, Conservation and Recycling*, 194, p. 107001.

Kocak, E. and Baglitas, H.H. (2022) 'The path to sustainable municipal solid waste management: Do human development, energy efficiency, and income inequality matter?', *Sustainable Development*, 30(6), pp. 1947–1962.

Kozlovskyi, S., Pasichnyi, M., Lavrov, R., Ivanyuta, N. and Nepytaliuk, A. (2020) 'An empirical study of the effects of demographic factors on economic growth in advanced and developing countries', *Comparative Economic Research. Central and Eastern Europe*, 23(3), pp. 45–68. https://doi.org/10.18778/1508-2008.23.03.

Kubanza, N.S. and Simatele, M.D. (2020) 'Sustainable solid waste management in developing countries: A study of institutional strengthening for solid waste management in Johannesburg, South Africa', *Journal of Environmental Planning and Management*, 63(2), pp. 175–188. https://doi.org/10.1080/09640568.2019.1576510.

Lenzen, M., Malik, A., Li, M., Fry, J., Weisz, H., Pichler, P.P., Chaves, L.S.M., Capon, A. and Pencheon, D. (2020) 'The environmental footprint of health care: A global assessment', *The Lancet Planetary Health*, 4(7), pp. e271–e279. https://doi.org/10.1016/S2542-5196(20)30121-2.

Maalouf, A. and Mavropoulos, A. (2023) 'Re-assessing global municipal solid waste generation', *Waste Management & Research*, 41(4), pp. 936–947.

Majeed, M.T. and Luni, T. (2020) 'Renewable energy, circular economy indicators and environmental quality: A global evidence of 131 countries with heterogeneous income groups', *Pakistan Journal of Commerce and Social Sciences (PJCSS)*, 14(4), pp. 866–912.

Mandpe, A., Paliya, S., Gedam, V.V., Patel, S., Tyagi, L. and Kumar, S. (2023) 'Circular economy approach for sustainable solid waste management', *Waste Management & Research*, 41(3), pp. 499–511.

Manisalidis, I., Stavropoulou, E., Stavropoulos, A. and Bezirtzoglou, E. (2020) 'Environmental and health impacts of air pollution: A review', *Frontiers in Public Health*, 8, Article 14. https://doi.org/10.3389/fpubh.2020.00014

Mehra, R., Kumar, S., Gupta, S., Saini, S., Singh, H. and Rajpoot, N.K. (2026) 'Advances in sustainable engineering and green technology: Integrating renewable systems, materials and digital innovation toward a circular economy', in *Advanced Computing and AI for Sustainable Environment, Energy, and Smart Manufacturing*. CRC Press, UK, pp. 50–54.

Melchor-Martínez, E.M., Macias-Garbett, R., Malacara-Becerra, A., Iqbal, H.M.N., Sosa-Hernández, J.E. and Parra-Saldívar, R. (2021) 'Environmental impact of emerging contaminants from battery waste: A mini review', *Case Studies in Chemical and Environmental Engineering*, 3, p. 100104. https://doi.org/10.1016/j.cscee.2021.100104.

Mihai, F.C. (2023) 'Circular economy and sustainable rural development', *Sustainability*, 15(3), p. 2139.

Mitra, S., Sarkar, A., Chauhan, A., Ghosh, A., Mohanty, S.K. and Bhattacharya, P. (2022) 'A review on environmental and health impacts of municipal solid waste management', *Environmental Science and Pollution Research*, 29(39), pp. 58520–58543.

Moraes, E.B., Kipper, L.M., Hackenhaar Kellermann, A.C., Austria, L., Leivas, P., Moraes, J.A.R. and Witczak, M. (2023) 'Integration of Industry 4.0 technologies with Education 4.0: Advantages for improvements in learning', *Interactive Technology and Smart Education*, 20(2), pp. 271–287.

Möslinger, M., Ulpiani, G. and Vetters, N. (2023) 'Circular economy and waste management to empower a climate-neutral urban future', *Journal of Cleaner Production*, 421, p. 138454.

Nguyen, C.P. and Su, T.D. (2022) 'The influences of government spending on energy poverty: Evidence from developing countries', *Energy*, 238, p. 121785. https://doi.org/10.1016/j.energy.2021.121785.

Nzediegwu, C. and Chang, S.X. (2020) 'Improper solid waste management increases potential for COVID-19 spread in developing countries', *Resources, Conservation and Recycling*, 161, p. 104947. https://doi.org/10.1016/j.resconrec.2020.104947.

Oloyede, A.A., Grace, D. and Faruk, N. (2023) 'Measuring the impact of the digital economy in developing countries: A systematic review and meta-analysis', *Heliyon*, 9(7), e17654. https://doi.org/10.1016/j.heliyon.2023.e17654.

Paes, M.X., de Oliveira, J.A.P., Mancini, S.D. and Rieradevall, J. (2024) 'Waste management intervention to boost circular economy and mitigate climate change in cities of developing countries: The case of Brazil', *Habitat International*, 143, p. 102990.

Pandya, D. (2024) *A Path to Formalization: Exploring the E-Waste Informal Sector's Integration Amid the Transition to Circular Economy* (Doctoral dissertation, Université d'Ottawal University of Ottawa).

Rai PK, Sonne C, Song H, Kim KH. The effects of COVID-19 transmission on environmental sustainability and human health: Paving the way to ensure its sustainable management. Sci Total Environ. 2022 Sep 10;838 (Pt 2):156039. doi: 10.1016/j.scitotenv.2022.156039.

Roy, A. (2021) *Demographics Unravelled: How Demographics Affect and Influence Every Aspect of Economics, Finance and Policy.* Hoboken, NJ: John Wiley & Sons.

Sahoo, A., Nanda, S., Mohanty, B., Nayak, A.K. and Bhattacharya, T. (2024) 'Insights into the management of food waste in developing countries with special reference to India', *Journal of Agriculture and Food Research*, 16, p. 101063. https://doi.org/10.1016/j.jafr.2023.101063.

Schroeder, P., Anggraeni, K. and Weber, U. (2019) 'The relevance of circular economy practices to sustainable development goals', *Journal of Industrial Ecology*, 23(1), pp. 77–95.

Sengupta, D., Ilankoon, I.M.S.K., Kang, K.D. and Chong, M.N. (2022) 'Circular economy and household e-waste management in India: Integration of formal and informal sectors', *Minerals Engineering*, 184, p. 107661.

Sharma, **K.D.** and Jain, **S.** (2020) 'Municipal solid waste generation, composition, and management: The global scenario', *Social Responsibility Journal*, 16(6), pp. 917–948.

Tafesse, S., Girma, Y.E. and Dessalegn, E. (2022) 'Analysis of the socio-economic and environmental impacts of construction waste and management practices', *Heliyon*, 8(3), e09169. https://doi.org/10.1016/j.heliyon.2022.e09169.

Tan, S.Y. and Taeihagh, A. (2020) 'Smart city governance in developing countries: A systematic literature review', *Sustainability*, 12(3), p. 899. https://doi.org/10.3390/su12030899.

Tisserant, A., Pauliuk, S., Merciai, S., Schmidt, J., Fry, J., Wood, R. and Tukker, A. (2017) 'Solid waste and the circular economy', *Journal of Industrial Ecology*, 21(3), pp. 628–640.

Turyahabwe, **R.**, Asaba, **J.**, Mulabbi, **A.**, Wamono, **E.** and Gudoyi, **P.M.** (2022) 'Environmental and socio-economic impact assessment of solid waste management practices in Mbale City, Uganda', *Ghana Journal of Geography*, 14(3), pp. 55–89.

UNEP (2021) *Global Methane Assessment.* Nairobi: United Nations Environment Programme.

United Nations Department of Economic and Social Affairs (UN DESA), Population Division (2019) *World Population Prospects 2019: Highlights.* New York: United Nations. Available at: UN DESA World Population Prospects 2019 (Accessed: 27 June 2026).

Valencia, A., Zhang, W. and Chang, N.B. (2022) 'Sustainability transitions of urban food-energy-water-waste infrastructure: A living laboratory approach for circular economy', *Resources, Conservation and Recycling*, 177, p. 105991.

Van Ewijk, S., Stegemann, J.A. and Ekins, P. (2018) 'Global Life cycle paper flows, recycling metrics, and material efficiency', *Journal of Industrial Ecology*, 22(4), pp. 686–693.

Van Fan, Y., Lee, C.T., Lim, J.S., Klemeš, J.J. and Le, P.T.K. (2019) 'Cross-disciplinary approaches towards smart, resilient and sustainable circular economy', *Journal of Cleaner Production*, 232, pp. 1482–1491.

Velis, C. (2017) 'Waste pickers in Global South: Informal recycling sector in a circular economy era', *Waste Management & Research*, 35(4), pp. 329–331.

Velis, C.A., Wilson, D.C., Gavish, Y., Grimes, S.M., Cheeseman, C.R. and Cottom, J.W. (2023) 'Socio-economic development drives solid waste management performance in cities: A global analysis using machine learning', *Science of the Total Environment*, 872, p. 161913. https://doi.org/10.1016/j.scitotenv.2023.161913.

Vinti, G., Bauza, V., Clasen, T., Medlicott, K., Tudor, T., Zurbrügg, C. and Vaccari, M. (2021) 'Municipal solid waste management and adverse health outcomes: A systematic review', *International Journal of Environmental Research and Public Health*, 18(8), p. 4331. https://doi.org/10.3390/ijerph18084331.

Wawrzyniak, D. and Doryń, W. (2020) 'Does the quality of institutions modify the economic growth–carbon dioxide emissions nexus? Evidence from a group of emerging and developing countries', *Economic Research-Ekonomska Istraživanja*, 33(1), pp. 124–144. https://doi.org/10.1080/1331677X.2019.1708770.

Weghmann, V. (2017) *Waste Management in Europe. Good Jobs in the Circular* Economy*?* Brussels: European Federation of Public Service Unions (EPSU). Report commissioned by EPSU.

Wikurendra, E.A., Csonka, A., Nagy, I. and Nurika, G. (2024) 'Urbanization and benefit of integration circular economy into waste management in Indonesia: A review', *Circular Economy and Sustainability*, 4(2), pp. 1219–1248.

Wilson, D.C., Velis, C. and Cheeseman, C. (2006) 'Role of the informal sector in waste management', *Habitat International*, 30(4), pp. 797–808.

Wilson, D.C., Velis, C. and Cheeseman, C. (2012) 'Role of informal sector recycling in waste management in developing countries', *Habitat International*, 30(4), pp. 797–808. https://doi.org/10.1016/j.habitatint.2005.09.005.

World Bank (2018) *What a Waste 2.0: A Global Snapshot of Solid Waste Management to 2050.* Washington, DC: World Bank.

World Bank (2020) *Poverty and Shared Prosperity 2020: Reversals of Fortune.* Washington, DC: World Bank. https://doi.org/10.1596/978-1-4648-1602-4.

World Bank (2022) *What Would It Take for Zambia's Copper Mining Industry to Achieve Its Potential?* Washington, DC: World Bank. Available at: World Bank Open Knowledge Repository (Accessed: 27 June 2026).

Yang, L., Chiu, Y.H., Chen, S., Chang, T.H. and Wang, Y. (2025) 'Estimating the efficiency of China's circular economy using an entropy approach', *Water Economics & Policy*, 11(2). https://doi.org/10.1142/S2382624X25500043

2 The Solid Waste Value Chain

2.1 GENERATION OF SOLID WASTE

2.1.1 Sources of Solid Waste

Solid waste results from various human, economic, and natural activities, presenting a complex challenge for municipalities globally. The primary sources consist of residential households generating food scraps, packaging, paper, plastics, textiles, and other discards; commercial establishments, including retail shops, offices, restaurants, and markets, contributing packaging materials, food residues, and paper; and institutional facilities such as schools, hospitals, and government offices producing medical waste, stationery paper, and canteen leftovers. Industrial activities contribute process residues, chemicals, metals, plastics, and by-products, whereas agricultural and mining operations produce crop residues, animal manure, fertilizers, pesticides, mine tailings, and slag. Moreover, construction and demolition sites generate rubble, concrete, timber, and metals. Diverse sources demonstrate the complexity of solid waste streams, with intensity varying based on population density, industrialization levels, and socioeconomic conditions.

Sharma and Jain (2020) highlight that the generation and composition of MSW are significantly affected by socioeconomic and demographic factors, with residential, commercial, and institutional sources contributing a substantial portion globally. The analysis indicates that urbanization and lifestyle changes, especially in developing countries, have intensified the generation of household and market waste, mirroring patterns seen in Lusaka. Chandrappa and Das (2024) further substantiate the categorization of solid waste sources by identifying primary generators such as households, industries, institutions, agriculture, and construction. Each source generates unique waste streams, including organic, recyclable, hazardous, and inert materials, necessitating specific management strategies. They emphasize that in the development of urban centers, household and market waste predominate, frequently surpassing the capacity of municipal collection and disposal systems. This evidence corroborates the observation made by the Lusaka City Council (2021) that household and market waste constitute most of the city's MSW. These studies collectively validate the categorization of waste sources and highlight the impact of socioeconomic factors on waste generation patterns in developing economies.

2.1.2 Types of Solid Waste

Solid waste can be classified based on its physical characteristics, source, and associated risk factors.

DOI: 10.1201/9781003591764-2

Biodegradable waste refers to organic materials, including food, paper, wood, and yard trimmings, that are capable of natural decomposition. This category prevails in developing economies, where food and agricultural residues constitute a significant portion of MSW (Godfrey et al., 2020). Figure 2.1 shows a picture of biodegradable waste.

Recyclable waste refers to materials including plastics, glass, metals, and paper that are capable of being recovered and reused within production systems. The potential for recycling is significant; however, the infrastructure for collection and processing is still inadequate in many regions of Africa (Sharma & Jain, 2020). Figure 2.2 shows a picture of recyclable waste.

Inert waste consists of construction debris, sand, stones, and demolition rubble that are predominantly nonreactive. The increase in waste streams is correlated with rapid urban expansion and infrastructure development in numerous cities (Chandrappa & Das, 2024). Figure 2.3 shows a picture of inert waste.

Hazardous waste includes toxic chemicals, batteries, medical waste, e-waste, and industrial residues, all of which present significant risks to human health and

FIGURE 2.1 Biodegradable waste.

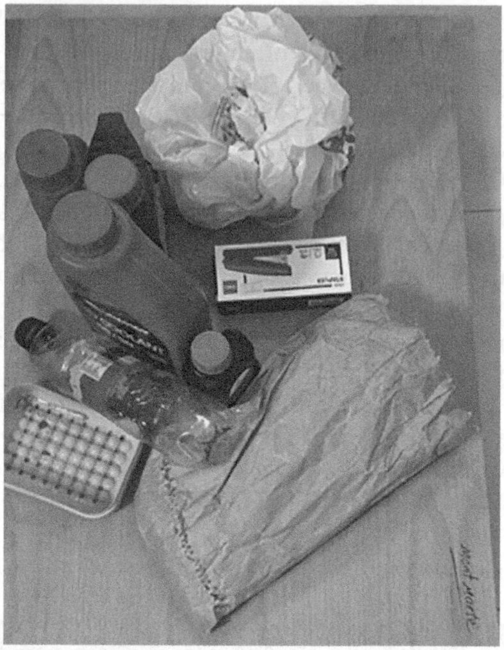

FIGURE 2.2 Recyclable waste.

FIGURE 2.3 Inert waste.

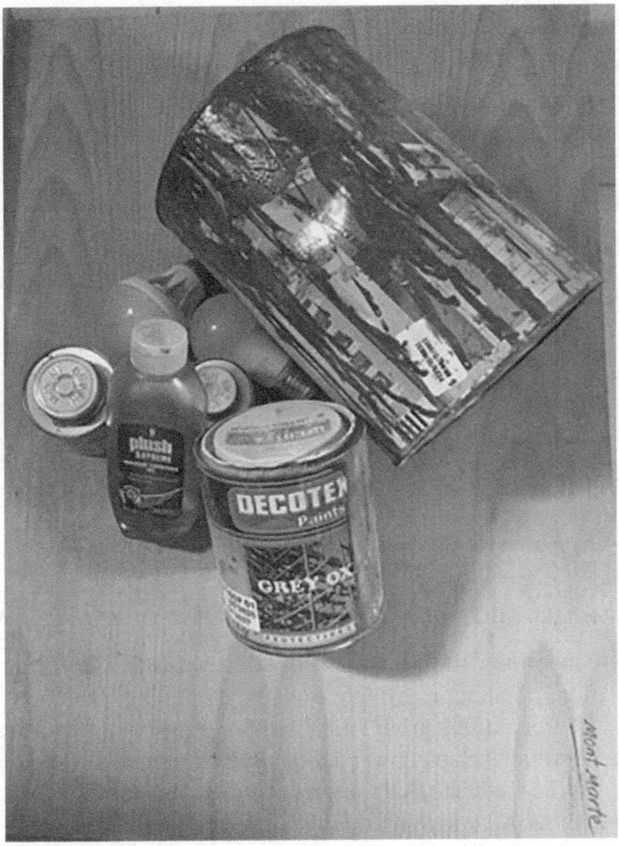

FIGURE 2.4 Hazardous waste.

the environment. The swift expansion of electronic products and healthcare services has led to an increase in hazardous waste volumes worldwide (Chen et al., 2020). Figure 2.4 shows a picture of hazardous waste.

Composite waste refers to materials such as textiles, rubber, and laminated products that present challenges in disassembly into individual components, thereby reducing their recyclability. In African contexts, composite waste is frequently disposed of in open dumps or landfills, attributed to insufficient separation technologies (Godfrey et al., 2020). Figure 2.5 shows a picture of composite waste.

The composition of waste streams typically mirrors consumer behavior, economic activities, and degrees of industrialization. The increasing demand for processed foods and beverages in developing economies has resulted in a notable rise in plastic packaging waste (Sharma & Jain, 2020; Chen et al., 2020). This underscores the necessity for sustainable consumption patterns and enhanced waste management strategies, especially in rapidly urbanizing areas.

FIGURE 2.5 Composite waste.

2.1.3 Characterization of Solid Waste in Developing Economies

In developing economies, the generation of solid waste is significantly affected by rapid urbanization, population growth, poverty dynamics, and changing consumption patterns. As urban areas grow and lifestyles transform, the characteristics and makeup of waste streams change, frequently placing pressure on municipal management systems. Studies indicate that these economies display unique characteristics in the composition of MSW, which markedly contrast with those observed in developed regions. Waste streams exhibit the following characteristics:

- The MSW in developing economies is primarily composed of biodegradable materials, including food scraps, paper, and yard trimmings. Research indicates that organic waste constitutes 50–70% of total waste, whereas in high-income countries, this figure ranges from 20% to 30% (Sharma & Jain, 2020; Krishna, 2024). Azam et al. (2020) conducted a case study in Lahore, Pakistan, which demonstrated that organic waste constituted approximately 60% of total municipal waste. This finding underscores a dual aspect: it presents a challenge regarding disposal and an opportunity for energy recovery via composting and biogas production.
- Recycling rates are low despite the availability of recyclable materials, including plastics, metals, and paper. This is attributed to insufficient infrastructure, ineffective policy enforcement, and inadequate market connections. Bui et al. (2022) identify substantial obstacles that emerging economies encounter in establishing effective recycling systems, such as financial limitations and technological deficiencies. Consequently, a significant portion of recyclable materials are deposited in uncontrolled dumpsites, exacerbating land and water pollution.
- The informal sector plays a significant role, with waste pickers, scrap dealers, and community-based organizations dominating recovery and recycling

activities in numerous cities across Africa and Asia. Their contributions significantly reduce the volume of materials sent to landfills; however, the work is frequently informal, inadequately regulated, and performed under unsafe conditions (Esteban et al., 2022). The informal sector plays a dual role; it is essential for material recovery yet remains unintegrated into formal waste management systems.

- Inadequate source segregation results in waste being collected in a mixed state, lacking prior separation into biodegradable, recyclable, and hazardous categories. This practice complicates recycling and treatment processes, thereby increasing costs and reducing efficiency (Ugwu et al., 2020). Research conducted on Nigerian university campuses indicated that unsorted waste streams restricted recovery options and resulted in increased disposal volumes, a situation that mirrors broader urban challenges.
- The increase in plastic waste is significant, driven by urbanization and a rise in the importation of packaged goods, making it a prominent element of waste streams. Chen et al. (2020) and Sharma and Jain (2020) emphasize the worldwide increase in plastics, whereas Godfrey et al. (2020) observe that inadequate management of plastic packaging in Africa significantly contributes to drainage obstruction, flooding, and environmental degradation. Bui et al. (2022) contend that without the adoption of CE strategies by emerging economies, the volumes of plastic waste will persist in increasing significantly.
- Lusaka, Zambia, produces more than 1,000 tons of MSW each day, with around 65% classified as biodegradable, 14% as plastics, 6% as paper, and the remaining portion consisting of metals, glass, and inert materials (ZEMA, 2020). This profile aligns with broader findings in developing economies, where waste is primarily organic but increasingly incorporates plastics and hazardous materials. Case studies from Africa, Asia, and Latin America highlight the necessity of implementing CE practices, including composting, waste-to-energy conversion, and the inclusion of informal actors, to mitigate the environmental and socioeconomic risks linked to unmanaged waste.

2.2 COLLECTION AND TRANSPORTATION

2.2.1 Methods of Waste Collection

Waste collection systems in developing economies exhibit diversity, influenced by socioeconomic conditions, infrastructural constraints, and policy frameworks. In contrast to developed nations, which predominantly utilize standardized systems like curbside and automated collection, developing countries employ a combination of formal and informal methods for waste management.

2.2.1.1 Curbside and Door-to-Door Collection

Curbside and door-to-door systems are considered effective in improving waste segregation and recovery, especially when integrated with public awareness initiatives. Azevedo et al. (2021) assert that insights from Germany demonstrate that organized curbside collection can enhance recycling rates and diminish reliance on landfills.

In developing countries, implementation is constrained by limited resources, inadequate institutional capacity, and substandard road infrastructure. Mwanza et al. (2018) observe that in Kitwe, Zambia, door-to-door collection is inconsistent and frequently augmented by informal waste pickers, which diminishes the efficiency of the waste management system. Jagun et al. (2023) assert that socioeconomic inequalities impact participation, as households with low income may lack the financial means for standardized bins or collection fees.

2.2.1.2 Community Bin Systems and Block Collection

Community bins are common in urban Africa and Asia due to their effectiveness in minimizing the logistical difficulties associated with servicing individual households. These systems are susceptible to overflow, illegal dumping, and unsanitary conditions, which compromise environmental health. Serge et al. emphasize that institutional deficiencies in Johannesburg hinder bin maintenance and prompt collection, leading to community opposition. Mwanza et al. (2018) argue that although block collection points in Kitwe serve as a temporary measure, they impede source separation, consequently diminishing the potential for resource recovery.

2.2.1.3 Drop-Off and Bring Systems

Drop-off systems, including eco-parks and centralized recycling stations, are infrequently observed in developing economies, attributed to limited public participation and inadequate enforcement mechanisms. Sharma and Jain (2020) demonstrate that, in the absence of financial incentives or robust awareness campaigns, households infrequently transport waste to designated collection points. Alzamora and Barros (2020) contend that charging models can influence participation positively or negatively, and in situations where affordability is a concern, such systems may exclude vulnerable populations.

2.2.1.4 Informal Sector Collection

The informal sector serves as a critical component of waste collection in numerous developing economies. Mwanza et al. (2018) emphasize the substantial role of informal waste pickers in resource recovery, frequently offsetting municipal inefficiencies. Jagun et al. (2023) emphasize that informal networks facilitate recycling and generate livelihoods for marginalized populations. Serge et al. (2020) note that the absence of institutional recognition places workers in unsafe conditions and hinders their integration into sustainable waste management frameworks.

2.2.1.5 Specialized Collection Systems

In many developing countries, specialized systems for managing hazardous, electronic, and medical waste remain underdeveloped. Sharma and Jain (2020) warn that inadequate management of these streams results in environmental pollution and health risks. Azevedo et al. (2021) propose that the implementation of customized collection frameworks from developed nations may improve safety; however, significant institutional and financial obstacles persist.

Developed economies predominantly utilize structured, mechanized, and policy-driven collection systems, whereas developing economies exhibit a hybrid scenario characterized by the coexistence of formal and informal systems. Mwanza et al. (2018) assert that the selection of a collection system significantly affects the potential for resource recovery and the implementation of a circular economy. Systemic challenges, including inadequate financing, weak governance, and limited citizen engagement, persistently undermine the efficiency of waste collection.

2.3 TRANSPORTATION MODELS AND CHALLENGES

2.3.1 Direct Haul Model

The direct haul model entails the transportation of waste directly from collection points to disposal sites, eliminating any intermediate stops. This practice is prevalent in small and medium-sized towns owing to its simplicity and minimal infrastructure needs. In numerous developing economies, this method proves unsustainable as urban areas grow and waste disposal sites are relocated further away (Abubakar et al., 2022). Extended travel durations, elevated fuel usage, and increased vehicle wear render direct hauling inefficient, particularly in densely populated urban areas. Chandrappa and Das (2024) assert that although direct haul is appropriate for smaller municipalities, it is seldom cost-effective in metropolitan areas. In contrast to developed economies, which benefit from advanced routing systems and modern fleets that facilitate direct haul, developing economies encounter constraints in technology and financial resources. The comparative analysis indicates that although direct haul is essential in resource-limited areas, shifting to transfer-based or hybrid systems is vital for long-term sustainability.

2.3.2 Station or Transfer-Based Model

The station or transfer-based model incorporates an intermediary phase in which waste is transported to transfer stations prior to bulk movement to remote treatment or disposal locations. This model decreases transportation expenses and facilitates intermediate sorting and compaction, thereby enhancing recycling prospects (Chandrappa & Das, 2024). In major urban areas, especially where disposal sites are distant, transfer stations are crucial for operational efficiency. Many developing economies face challenges in establishing and maintaining infrastructure due to capital constraints and weak institutional frameworks (Kubanza & Simatele, 2020). Inadequate planning and enforcement lead to the underutilization or absence of transfer stations, resulting in a dependence on direct haul methods. In developed economies, transfer stations are commonly implemented, facilitated by sophisticated compaction technologies and well-regulated logistics systems. The comparative analysis reveals disparities in financial investment and institutional capacity, indicating that although transfer-based models are optimal, developing economies necessitate customized, cost-effective alternatives for effective implementation.

2.3.3 Decentralized/Community-Based Model

The decentralized transport model is commonly utilized in informal settlements and peri-urban regions, where traditional municipal services are inadequate.

Communities or cooperatives utilize small vehicles, including carts, tricycles, or pickup trucks, for local waste transport management. This model improves accessibility in narrow streets and creates employment opportunities; however, its limited capacity and absence of formal recognition diminish its efficiency. Municipal integration frequently exhibits deficiencies, resulting in fragmented services and variable coverage. Although effective in addressing the needs of underserved areas, community-based models are insufficient for managing the increasing volumes of urban waste independently. In developed economies, decentralized collection is infrequent, as formal municipal systems generally guarantee universal access, bolstered by advanced technologies. Comparative analysis indicates that community-based models serve as essential stopgaps in developing economies. Their potential is enhanced through integration with formal systems, which can improve coverage and promote inclusive, hybrid waste transport solutions that balance accessibility and efficiency.

2.3.4 Private Sector and Informal Model

Private sector entities and informal waste pickers are essential in addressing service deficiencies in contexts where municipal systems lack adequate funding or efficiency. These groups frequently utilize small vehicles or contract-based trucks for waste transportation, providing flexible and demand-responsive services (Abubakar et al., 2022). Nonetheless, their operations often lack regulation, resulting in illegal dumping, inadequate accountability, and hazardous conditions. Kubanza and Simatele (2020) contend that the incorporation of these actors into formal systems via regulation, training, and partnerships would enhance efficiency and accountability. Developing economies depend significantly on private and informal entities because of institutional deficiencies, whereas developed economies manage private involvement through contractual agreements, performance assessments, and compliance regulations. The comparative analysis indicates that while private and informal systems improve accessibility in resource-limited environments, they need to evolve into regulated partnerships to ensure sustainability, drawing lessons from organized private sector involvement in developed contexts.

2.3.5 Intermodal Model

The intermodal model incorporates various transportation modes, including road, rail, and waterways, to enhance the efficiency of long-distance waste transport. This system demonstrates high efficiency in bulk waste movement, leading to reductions in both costs and environmental impacts (Chandrappa & Das, 2024). However, in developing economies, poor infrastructure and limited investment in rail or waterways hinder adoption. Bryan et al. (2020) observe that inadequate urban infrastructure in the Global South significantly limits the implementation of advanced intermodal strategies. In developed economies, intermodal transport is more viable owing to strong infrastructure networks and institutional backing, frequently utilized to connect urban areas with distant treatment facilities or landfills. Comparative analysis demonstrates that intermodal systems, although aspirational in developing economies, present a significant opportunity for infrastructure investment to improve sustainability and efficiency, thereby connecting local waste generation with distant treatment capacity.

2.4 CHALLENGES IN WASTE TRANSPORTATION

2.4.1 Infrastructure Limitations

Infrastructure limitations represent a major obstacle to effective waste transportation in developing economies. Poor road conditions, especially in peri-urban and informal settlements, frequently hinder access to households, while seasonal rains render unpaved roads impassable (Bryan et al., 2020). This leads to missed collections and a rise in illegal dumping. Vaverková et al. (2021) emphasize that COVID-19 revealed existing vulnerabilities, exacerbating disruptions in transport systems within already constrained environments. Conversely, developed economies experience advantages due to superior urban planning and road infrastructure, which improve reliability and coverage. The comparative analysis indicates that developed economies enhance routing and fleet performance, whereas developing countries continue to face fundamental infrastructure deficiencies that hinder the establishment of efficient transport systems.

2.4.2 High Operational Costs

Waste transportation represents a significant portion of municipal budgets, frequently comprising 40–60% of total waste management expenses in developing economies (Abubakar et al., 2022). Increasing fuel costs, aging fleets, and elevated labor expenses contribute to inefficiencies, hindering municipalities' ability to maintain consistent collection schedules. Chandrappa and Das (2024) assert that routing optimization and investment in modern fleets are essential to achieving sustainable operational costs. Conversely, developed economies achieve cost reductions via advanced logistics, automation, and energy-efficient fleets. The comparative analysis indicates that operational expenses are a global issue; however, developed economies mitigate these costs via technology and economies of scale, whereas developing countries are constrained by cost-intensive, low-efficiency cycles.

2.4.3 Limited Transfer Stations

Transfer stations play a crucial role in minimizing haul distances and facilitating efficient bulk transport; however, numerous developing economies are deficient in such infrastructure due to financial and institutional limitations (Kubanza & Simatele, 2020). Where present, these facilities are frequently inadequately maintained or underused, resulting in ongoing dependence on expensive direct hauls. The lack of transfer facilities restricts opportunities for waste sorting and recycling during an intermediate phase. In developed economies, transfer stations are prevalent, underpinned by automation, compaction technologies, and robust institutional frameworks.

The comparative analysis reveals a significant disparity: developed countries enhance transfer infrastructure for efficiency and CE results, whereas developing economies fall short due to insufficient investment and planning capabilities.

2.4.4 Weak Integration of Informal Sector

In developing economies, informal actors deliver crucial transport services, particularly in underserved communities, yet they are often excluded from formal planning processes. Abubakar et al. (2022) emphasize that their contributions to service coverage and recycling are significant; however, the absence of recognition leads to fragmented and unsafe practices. Kubanza and Simatele (2020) assert that institutional integration via training and recognition may enhance accountability. Conversely, developed economies seldom depend on informal actors due to the professionalization and regulation of services. The comparative analysis indicates that although the informal sector is essential in developing economies, it highlights a deficiency in inclusivity, which could be addressed through the implementation of hybrid formal–informal collaboration frameworks.

2.4.5 Environmental and Health Concerns

Inadequately managed waste transportation presents significant environmental and health hazards in developing economies. Open or poorly covered trucks frequently release waste, resulting in litter, leachate, and odor, while workers are exposed to hazardous materials without appropriate protective equipment (Chandrappa & Das, 2024). Vaverková et al. (2021) emphasize that during the COVID-19 pandemic, improper management of medical and household waste exacerbated environmental and health risks in numerous cities in the Global South. In advanced economies, stringent regulations, specialized vehicles, and occupational safety standards effectively reduce these risks. The comparative analysis indicates that developing economies experience greater environmental and health challenges from waste transport, attributed to insufficient enforcement and inadequate equipment, whereas developed countries attain safer results through effective regulation and advanced technology.

2.4.6 Institutional and Financial Barriers

Institutional and financial obstacles substantially impede efficient waste transportation. Weak governance structures, fragmented policies, and insufficient sustainable financing constrain investment in fleets, transfer stations, and logistics planning (Kubanza & Simatele, 2020). Abubakar et al. (2022) argue that institutional inefficiencies perpetuate financial deficits, creating a cycle of underperformance. Corruption and mismanagement intensify resource deficiencies. In contrast, developed economies gain advantages from stable institutions, effective regulatory enforcement, and diverse financing models, including public–private partnerships. The comparative analysis reveals that global financial challenges are exacerbated in developing economies due to institutional weaknesses, underscoring the necessity for governance reforms to enhance sustainability.

2.4.7 Urbanization and Increasing Waste Volumes

Rapid urbanization in developing economies produces increasing waste volumes that exceed the capacity of current transport systems. Bryan et al. (2020) contend that urban areas in the Global South grow at a rate that outpaces infrastructure investment, resulting in backlogs and instances of illegal dumping. Debrah et al. (2021) assert that public awareness and education are crucial for the sustainable management of this growth. In developed economies, improved infrastructure planning and technological systems facilitate the alignment of waste volumes with transport capacity. The comparative analysis reveals a significant divergence: developed countries implement scalable systems to accommodate urban growth, whereas developing economies experience an increasing disparity between waste generation and transport capacity, indicating a pressing need for reforms and investments.

Transportation plays a central role in SWM, linking collection systems with treatment and disposal facilities. In developing economies, a range of models exist, each with distinct benefits and limitations. The direct haul model remains common due to its simplicity, though it becomes inefficient for long distances. The transfer-based model offers cost savings and recycling opportunities but demands significant capital and institutional strength, which many cities lack. Decentralized and community-based models improve access in informal settlements and create local jobs, yet they remain fragmented without municipal integration. The private sector and informal systems fill critical gaps, though they often operate unregulated, leading to risks of illegal dumping. Finally, the intermodal model is highly efficient in developed economies but remains aspirational in much of the Global South due to infrastructure deficits.

These models are shaped by multiple transportation challenges. Infrastructure limitations such as poor road networks and seasonal inaccessibility undermine reliability, while high operational costs consume up to 60% of municipal budgets. The absence of transfer stations further forces costly direct hauls, reducing opportunities for waste recovery. Weak integration of informal actors leaves essential contributors unrecognized, while environmental and health risks are amplified by open trucks and poor handling practices. Institutional and financial barriers, including weak governance and underfunding, exacerbate inefficiencies, particularly as urbanization rapidly increases waste volumes beyond transport capacity.

Overall, the comparative picture shows that while developed economies deploy well-regulated, technologically advanced systems, developing countries adopt hybrid and adaptive models constrained by financial, institutional, and infrastructural realities. The path forward lies in strengthening governance, investing in infrastructure such as transfer stations, integrating informal and private actors, and adopting scalable transport strategies to meet the demands of fast-growing urban populations.

2.5 ROLE OF INFORMAL SECTOR IN WASTE COLLECTION

Management of MSW constitutes a significant urban challenge in developing economies, primarily attributed to constrained institutional capacity, swift urbanization, and insufficient financial resources. The informal sector has become a significant participant in waste collection and recycling. Informal waste pickers, small-scale collectors, and community-based organizations frequently offer services in areas where municipal systems are inadequate, effectively addressing significant gaps in waste collection and recovery. Researchers have acknowledged that comprehending and incorporating the informal sector's role is essential for effective SWM and the attainment of CE objectives (Tong et al., 2021; Harfadli et al., 2024).

2.5.1 Contribution to Waste Collection and Recycling

The informal sector plays a significant role in primary collection, particularly in low-income settlements and peri-urban regions that lack adequate municipal services. Research conducted in Vietnam indicates that informal collectors play a crucial role in door-to-door collection and recycling, managing waste streams that would otherwise go uncollected (Tong et al., 2021). In Indonesia, informal waste pickers significantly alleviate municipal burdens by diverting recyclables prior to their arrival at dumpsites (Satori et al., 2020).

Kubanza and Simatele (2020) illustrate the role of informal collectors in South Africa as they address deficiencies in the institutional framework by providing essential services in Johannesburg's waste collection. Their contribution diverts waste from overflowing landfills and generates livelihoods for marginalized groups. It is estimated that informal actors recover between 50% and 80% of recyclables in numerous developing cities worldwide (Harfadli et al., 2024). This illustrates their essential function in resource recovery and in promoting the wider CE initiative.

2.5.2 Livelihoods and Socioeconomic Significance

In addition to environmental advantages, the informal sector offers essential socioeconomic opportunities. Waste picking provides income for many urban poor, frequently encompassing women, migrants, and youth who have restricted access to formal employment opportunities. The informal sector in Indonesia serves as a livelihood safety net, offering subsistence for households marginalized from formal labor markets (Satori et al., 2020).

In India, informal e-waste collectors play a crucial role in household-level collection and recycling, positioning them as key players in the transition to a CE in electronics (Sengupta et al., 2022). Their activities mitigate unemployment pressures and enhance urban resilience. In Zambia, research emphasizes initiatives like Manja Pamodzi, which enables informal collectors to provide recyclable materials to industry, thereby generating employment and advancing national sustainability goals (Mwanza et al., 2017).

2.5.3 Environmental Impact

The informal sector plays a crucial role in promoting environmental sustainability through waste diversion from landfills, reduction of open dumping, and mitigation of greenhouse gas emissions. Tong et al. (2021) assert that informal recycling in

Vietnam mitigates the environmental impact of waste systems by reclaiming plastics, paper, and metals, which would otherwise produce methane and leachates.

Sengupta et al. (2022) highlight that informal e-waste recycling in India is crucial for mitigating electronic pollution, although it faces challenges including unsafe practices and insufficient protective equipment. The informal sector contributes to cleaner urban environments by decreasing the volume of waste directed towards uncontrolled disposal, thereby aligning with the United Nations Sustainable Development Goals, specifically SDG 11 and SDG 12.

2.5.4 Challenges Facing the Informal Sector

The informal sector, while contributing significantly, functions under precarious conditions. Harfadli et al. (2024) identify multiple challenges, such as unsafe working conditions, insufficient social protection, and exposure to hazardous substances. Informal workers frequently engage in waste collection without protective equipment, rendering them susceptible to injuries and illnesses.

Institutional neglect exacerbates these challenges. Kubanza and Simatele (2020) observe that in South Africa, weak governance and insufficient institutional support marginalize informal collectors, resulting in tensions between formal and informal systems. Sengupta et al. (2022) assert that unregulated e-waste management in India exposes workers to hazardous substances, compromising health and safety and emphasizing the necessity for policy interventions.

2.5.5 Pathways for Integration and Formalization

Due to their essential contributions, researchers support the formalization and integration of informal workers into municipal frameworks. Strategies encompass the formation of cooperatives, the establishment of public–private partnerships (PPPs), and the integration of extended producer responsibility (EPR) schemes. In Brazil, waste picker cooperatives illustrate the potential for formal recognition of informal workers, provision of infrastructure, and integration into municipal recycling systems (Harfadli et al., 2024).

Mwanza et al. (2017) contend that the integration of informal collectors into formalized systems in Zambia necessitates policy alignment, stakeholder engagement, and infrastructural investment. The engineering management perspective highlights that integration increases efficiency, enhances occupational health, and guarantees adherence to national sustainability objectives. Sengupta et al. (2022) emphasize that integrating formal and informal systems in India's e-waste sector can lead to hybrid models that enhance efficiency while promoting social inclusion.

Comparative evidence indicates both similarities and context-specific variations in the function of informal sectors within developing economies. Informal collectors in Vietnam operate within a fragmented system and play a crucial role in sustainability (Tong et al., 2021). In Indonesia, their function is socioeconomic, serving as a survival strategy for impoverished households (Satori et al., 2020). Informal collectors in South Africa play a vital role yet encounter institutional exclusion

(Kubanza & Simatele, 2020). India exemplifies the dominance of informal actors in e-waste collection; however, unsafe practices hinder sustainable integration (Sengupta et al., 2022).

The Manja Pamodzi initiative in Zambia exemplifies how collaborations between industry and informal collectors can enhance circularity and support livelihoods (Mwanza et al., 2017). These cases highlight the importance of informal actors in waste systems, necessitating context-specific strategies for their formalization and empowerment.

2.6 PROCESSING AND TREATMENT

2.6.1 Methods for Waste Segregation and Processing

2.6.1.1 Segregation at Source

Source segregation is acknowledged as a fundamental aspect of effective SWM; however, its implementation is inadequate in many developing economies. Households and businesses frequently exhibit reluctance or inability to segregate organic, recyclable, and hazardous wastes, attributed to insufficient awareness, inadequate infrastructure, and ineffective regulatory enforcement (Kumari & Raghubanshi, 2023). In numerous cities across Africa and Asia, municipal authorities offer a single collection bin, leading to mixed waste streams that hinder downstream processing (Sharma & Jain, 2020). Successful pilot projects illustrate the potential of source segregation. In South Africa, the implementation of awareness campaigns alongside the introduction of color-coded bins in specific neighborhoods has notably enhanced recycling rates (Godfrey et al., 2020). During the COVID-19 pandemic in India, local governments encouraged the segregation of infectious waste at the household level, demonstrating that behavior change can be accomplished in urgent circumstances (Ganguly & Chakraborty, 2021). Notwithstanding these achievements, the challenge of scaling up persists. Jagun et al. (2023) highlight that community engagement and incentives, including pay-as-you-throw schemes or rewards for recyclables, can enhance household participation. Segregation at source is essential for enhancing waste systems; however, it necessitates strong institutional backing, ongoing education, and socioeconomic incentives to be effective in developing contexts.

2.6.1.2 Informal Sector Sorting

In many developing economies, the lack of robust formal systems results in informal waste pickers and community-based groups primarily conducting segregation activities. Actors gather recyclables from households, streets, and dumpsites, providing industries with inexpensive secondary raw materials (Kubanza & Simatele, 2020). Their contribution is significant, achieving recovery rates of 50–80% of recyclables in certain cities (Azevedo et al., 2021). In Johannesburg, South Africa, informal collectors alleviate the burden on municipalities by offering door-to-door recovery services, yet they are excluded from official planning frameworks (Kubanza & Simatele, 2020). In India, informal actors are pivotal in the recycling of plastics, paper, and metals, mitigating the shortcomings of formal waste management systems during the COVID-19 pandemic (Ganguly & Chakraborty, 2021). Nonetheless, their operations exhibit inadequate working conditions, insufficient social protection, and exposure to health risks (Jagun et al., 2023). Integrating these workers into formal systems through cooperatives or public–private partnerships has proven effective in countries like Brazil, where waste picker organizations receive municipal contracts to provide collection services (Azevedo et al., 2021). Recognizing and formalizing the role of informal actors enhances segregation efficiency and fosters opportunities for social inclusion and economic empowerment in waste management.

2.6.1.3 Centralized Sorting and Recycling Facilities

Centralized sorting at community-based transfer stations or materials recovery facilities (MRFs) offers a systematic method for segregation and recycling in developing nations. These facilities process mixed waste sourced from households and utilize a workforce that typically includes both formal employees and informal pickers to extract valuable materials. Kumari and Raghubanshi (2023) assert that these facilities improve efficiency by consolidating recyclables for industries and decreasing reliance on landfills. In South Africa, materials recovery facilities (MRFs) backed by municipalities and nongovernmental organizations (NGOs) have illustrated the effectiveness of partnerships in establishing sustainable recycling models (Godfrey et al., 2020). In Nigeria and Ethiopia, donor-funded sorting facilities have been established; however, their long-term sustainability is compromised by inadequate funding and insufficient market linkages (Jagun et al., 2023). Azevedo et al. (2021) contend that insights from Germany's dual system, characterized by a strong centralized recycling infrastructure, can be applied to developing contexts through the integration of community cooperatives and informal actors. Nevertheless, in the absence of dependable waste collection logistics, these facilities continue to be underutilized. Sharma and Jain (2020) emphasize that centralized sorting requires robust policy frameworks, including extended producer responsibility (EPR), to ensure adequate financing and a stable waste supply. MRFs show potential but necessitate stable governance and financial frameworks for success in developing economies.

2.6.1.4 Composting

Composting represents a practical and extensively advocated method for waste processing in developing economies, attributed to the significant percentage of organic matter in MSW, typically ranging from 50% to 70% (Sharma & Jain, 2020). Converting biodegradable waste into nutrient-rich compost alleviates landfill pressure and offers a sustainable alternative to chemical fertilizers. Small-scale community composting in India has facilitated peri-urban farming systems by providing low-cost soil amendments, thus connecting waste management to food security (Kumari & Raghubanshi, 2023). In Ethiopia and Nigeria, donor-funded composting projects have been implemented to enhance soil fertility and decrease methane emissions from dumpsites (Godfrey et al., 2020). Challenges remain, such as limited public awareness, insufficient space for composting facilities in urban settings, and weak market demand for compost relative to synthetic fertilizers (Jagun et al., 2023). Social attitudes significantly affect adoption, as households frequently resist separating organic waste, thereby diminishing feedstock quality (Ganguly & Chakraborty, 2021). When integrated with educational initiatives and local incentives, composting offers a cost-effective and scalable approach to organic waste management in developing economies, facilitating transitions to CE by converting waste into a resource for sustainable agriculture.

2.6.1.5 Anaerobic Digestion

Anaerobic digestion (AD) represents a significant approach for the management of the substantial organic fraction of MSW in developing economies. AD decreases waste volumes and produces biogas, which can be utilized for cooking, electricity, and heating, thereby providing advantages in both waste management and renewable energy generation (Sharma & Jain, 2020). Pilot biogas plants in South Africa have shown the potential of anaerobic digestion to meet local energy demands, especially in peri-urban and rural areas (Godfrey et al., 2020). In Ethiopia, small-scale digesters facilitated by development agencies have allowed farmers to utilize both energy and digestate fertilizer, demonstrating the role of anaerobic digestion in improving agricultural productivity (Kumari & Raghubanshi, 2023). Nonetheless, adoption is constrained by technical and financial limitations. High capital costs, a shortage of skilled technicians, and inconsistent waste segregation practices impede the expansion of AD projects (Jagun et al., 2023). Additionally, numerous facilities encounter operational challenges, such as feedstock variability and system failures. Research indicates that, despite existing barriers, appropriate policy support, subsidies, and integration into urban energy strategies can enable AD to serve as a crucial element of sustainable waste management in developing economies, aligning waste treatment with clean energy and climate mitigation goals (Ganguly & Chakraborty, 2021).

2.6.1.6 Waste-to-Energy

Waste-to-energy (WtE) technologies, including incineration and gasification, are gaining attention as potential solutions for the increasing waste volumes in developing economies. WtE has the potential to decrease reliance on landfills and produce electricity; however, its viability in low- and middle-income countries is frequently debated. The main challenge is the composition of waste; unsorted, moisture-laden streams with high organic content are not conducive to efficient incineration (Sharma & Jain, 2020). In India, numerous WtE plants have faced challenges stemming from variable feedstock quality and public resistance related to environmental issues (Ganguly & Chakraborty, 2021). In Johannesburg, South Africa, efforts to establish WtE facilities have encountered opposition stemming from concerns regarding health risks and affordability (Kubanza & Simatele, 2020). Successful cases exist in which adapted technologies, including presorting and mechanical–biological treatment, have enhanced the viability of WtE processes. Azevedo et al. (2021) indicate that insights from Germany's experience, characterized by robust regulation and elevated segregation rates, can guide developing economies in establishing effective WtE systems. Without substantial investments, effective governance, and stringent environmental standards, WtE may become financially unviable and socially contentious. Although promising, WtE in developing economies requires careful adaptation to local waste profiles and should be integrated with broader CE strategies instead of being pursued as an isolated solution.

Composting, AD, and WtE technologies present unique approaches to the challenges of MSW management in developing economies; however, their effectiveness and sustainability are significantly influenced by local contexts.

Composting represents a highly adaptable solution due to the significant organic content in waste streams of numerous low- and middle-income countries, typically ranging from 50% to 80% (Sharma & Jain, 2020). This approach is economically efficient and straightforward to execute, and directly benefits agricultural systems through the production of organic fertilizer. Nonetheless, its efficacy is constrained by inadequate segregation practices and diminished market demand for compost relative to synthetic fertilizers (Kumari & Raghubanshi, 2023).

AD offers significant technological advantages by generating renewable energy and organic fertilizer, thereby serving as a dual solution approach. Pilot projects in South Africa and Ethiopia have demonstrated potential, with small-scale digesters facilitating energy access in rural and peri-urban communities (Godfrey et al., 2020). High capital costs, operational complexity, and a shortage of skilled technicians limit its scalability (Jagun et al., 2023).

Conversely, WtE provides extensive solutions through the conversion of waste into electricity, supporting urban energy strategies and minimizing dependence on landfills. However, its implementation in developing economies is significantly limited by inadequate waste segregation, elevated organic content, and public resistance to incineration (Kubanza & Simatele, 2020; Ganguly & Chakraborty, 2021). Models from Europe, especially Germany, demonstrate that WtE necessitates sophisticated infrastructure, strong regulatory frameworks, and effectively segregated waste, conditions frequently lacking in developing regions (Azevedo et al., 2021).

Composting and AD offer more immediate and contextually relevant solutions for waste management in developing economies, whereas WtE remains aspirational and requires significant capital investment. The future direction involves the integration of these methods into hybrid systems, emphasizing the scaling of composting and AD for organic waste, while selectively exploring WtE in urban centers where conditions permit. This integrated approach enhances environmental sustainability and aligns with CE principles and Sustainable Development Goals.

2.7 TECHNOLOGIES FOR WASTE TREATMENT

2.7.1 Composting

Composting represents a highly effective treatment technology in developing economies, as MSW is predominantly composed of biodegradable materials, which often account for 50–70% of the total waste generated (Nanda & Berruti, 2021). Converting organic waste into compost enables cities to decrease landfill waste volume and supply affordable fertilizer for agricultural applications. Community-scale windrow composting has proven effective in countries like India, and donor-supported initiatives in Ethiopia have enhanced soil fertility and mitigated methane emissions from unmanaged waste (Hammed & Sridhar, 2020). Scaling composting encounters obstacles. Inadequate source segregation results in contamination, and the compost market is adversely affected by the availability of less expensive synthetic fertilizers

(Azevedo et al., 2021). Urban space limitations and challenges in odor management exacerbate operational complexities. Evidence from developed nations indicates that robust regulation, established standards, and market incentives enhance compost adoption. Composting continues to be a cost-effective and environmentally sustainable method for waste treatment in low- and middle-income countries, particularly when combined with awareness campaigns, incentive programs, and policies that support organic agriculture and CE connections.

2.7.2 Anaerobic Digestion

AD represents a sophisticated biological treatment method that effectively tackles waste reduction and fulfills renewable energy requirements. This process entails the anaerobic decomposition of organic waste, resulting in the generation of biogas and nutrient-dense digestate. The dual benefit of AD renders it particularly appealing in developing economies, where there is a significant presence of organic waste alongside energy shortages (Nanda & Berruti, 2021). Pilot AD projects in South Africa and Ethiopia illustrate the dual capability of digesters to produce electricity and supply organic fertilizer for agricultural use (Godfrey et al., 2020). AD effectively aligns with CE objectives by integrating waste management with energy and food systems. Nonetheless, adoption is still constrained. AD systems necessitate significant capital investment and demand skilled operation and maintenance, which are frequently deficient in low-resource settings (Hammed & Sridhar, 2020). Contamination of feedstock due to inadequate waste segregation diminishes efficiency. Farooq et al. (2021) highlight the necessity of multi-criteria frameworks for assessing feasibility, considering technical, financial, and institutional readiness. Although promising, AD requires careful adaptation to local contexts, supported by subsidies, technical training, and integration with renewable energy policies. In the long term, AD has the potential to function as a scalable solution for sustainable waste treatment, contingent upon robust governance and financing mechanisms.***

2.7.3 Waste-to-Energy

WtE technologies such as incineration, gasification, and refuse-derived fuel (RDF) are increasingly promoted to reduce landfill reliance while generating electricity. In principle, these technologies offer a high-volume solution to urban waste challenges. However, their success depends heavily on waste composition and institutional capacity. In most developing economies, municipal waste is unsorted and moisture-heavy, dominated by organics, which undermines thermal efficiency and increases emissions control costs (Nanda & Berruti, 2021). Experiences from India show that several WtE plants have struggled or failed due to feedstock inconsistency and public resistance over environmental concerns (Ganguly & Chakraborty, 2021). In Johannesburg, South Africa, attempts to establish incineration projects have faced affordability and governance challenges (Kubanza & Simatele, 2020). Still, with pre-sorting, drying, and strong regulatory frameworks, WtE can play a complementary role. Khan and Kabir (2020) highlight that multi-criteria assessments are essential to ensure economic, environmental, and social sustainability before implementation. Lessons from Germany show that successful WtE depends on high segregation rates, extended producer responsibility, and robust monitoring (Azevedo et al., 2021). Thus, while WtE remains attractive in theory, in practice it must be selectively applied in urban centers with strong governance capacity, financial resources, and well-structured waste streams.

2.7.4 Recycling and Materials Recovery Facilities

Recycling and MRFs play a crucial role in extracting value from waste streams in developing economies. Mechanical and manual sorting processes facilitate the recovery of plastics, metals, glass, and paper, supplying raw materials for local industries. MRFs vary from basic, community-managed sorting centers to sophisticated facilities equipped with optical sensors and conveyor systems (Cheah et al., 2022). In nations such as South Africa and Nigeria, donor- and NGO-supported MRFs have enhanced recycling outcomes; however, their sustainability is frequently compromised by inadequate financing and irregular availability of clean recyclables (Godfrey et al., 2020). Informal waste pickers frequently surpass machinery in the recovery of recyclables, especially plastics, owing to their adaptability and cost-effectiveness (Kubanza & Simatele, 2020). These workers function in substandard conditions and lack acknowledgment. Browning et al. (2021) highlight that contaminated plastics and mixed waste streams diminish recycling efficiency, underscoring the necessity for upstream segregation and policy measures like extended producer responsibility. Azevedo et al. (2021) identify Germany's dual system as a model, wherein the separation of households and financing from industry contribute to the stabilization of recycling markets. Hybrid systems that incorporate informal actors alongside formal MRFs, bolstered by incentives and awareness initiatives, offer a viable approach for enhancing recycling rates and quality in developing economies.

2.7.5 Digital and Integrated Technologies

The future of waste treatment in developing economies depends on the integration of digital tools, smart systems, and robust institutional governance. Innovations associated with Industry 4.0, such as sensor-enabled bins, AI-powered sorting machines, and blockchain-based traceability, are initiating a transformation in waste management towards a data-driven CE model (Cheah et al., 2022). These technologies enhance efficiency, minimize contamination, and foster transparency within recycling chains. Their adoption in developing contexts is constrained by elevated costs, infrastructure deficiencies, and the necessity for technical expertise. Institutional strengthening is a necessary condition. Case studies from Johannesburg indicate that inadequate planning and governance can compromise the effectiveness of even well-intentioned projects (Kubanza & Simatele, 2020). Lessons from Germany indicate that regulatory frameworks, financing mechanisms, and extended producer responsibility are crucial for stabilizing digital and high-tech interventions (Azevedo et al., 2021). Farooq et al. (2021) assert that decision-making frameworks must direct technology selection to ensure compatibility with local socio-economic conditions. Integrating informal actors into digitalized systems enhances inclusivity and improves material recovery. Digital and integrated technologies should not be viewed as isolated solutions; rather, they need to be incorporated into governance frameworks, financial models, and participatory planning to guarantee long-term sustainability. The integration of these approaches results in resilient, transparent, and inclusive waste management systems.

2.7.6 Comparative Analysis

Composting, AD, WtE, recycling via MRFs, and digital integration present distinct opportunities and limitations for waste management in developing economies. Composting is the most accessible and scalable solution, effectively addressing waste streams primarily composed of organic materials. It is cost-effective, environmentally sustainable, and directly connected to agriculture; however, it necessitates market development and improved segregation practices for success. AD provides additional benefits by producing renewable energy in conjunction with fertilizer, representing a circular solution; however, its implementation is limited by significant capital expenditures, technical intricacies, and the necessity for dependable feedstock.

WtE technologies offer significant potential for large-scale energy recovery and decreased reliance on landfills; however, they are less effective for mixed, moisture-laden waste typically found in developing regions. Success is contingent upon effective governance, stringent regulation, and efficient pre-sorting systems, which are frequently inadequate. Recycling and MRFs are essential for the recovery of resources, especially plastics and metals. Nonetheless, their efficacy is undermined by contamination and insufficient funding, with informal actors frequently addressing essential deficiencies. Digital and integrated technologies indicate a forward-looking approach, providing efficiency and transparency via data-driven systems. However, their elevated costs and technical prerequisites render them viable solely when integrated with institutional enhancement and inclusive governance.

Biological methods, including composting and AD, are the most context-appropriate strategies for the short to medium term. In contrast, WtE and digital technologies necessitate long-term investments in governance, infrastructure, and financing. Recycling and MRFs are pivotal, necessitating the integration of informal and formal systems. Developing economies require hybrid, context-sensitive strategies that integrate low-cost, accessible technologies with the gradual implementation of advanced systems. This equilibrium promotes environmental sustainability, social inclusion, and coherence with CE principles and sustainable development objectives.

2.8 CASE STUDIES FROM DEVELOPING ECONOMIES

2.8.1 India—Decentralized Composting Anchored by Source Segregation

Decentralized composting in India has emerged as an effective waste treatment method, primarily owing to the prevalence of biodegradable materials in MSW. Community-scale initiatives, underpinned by source segregation, demonstrate that composting effectively decreases the quantity of waste directed to dumpsites while producing marketable organic fertilizer. In Vadodara, a women's self-help group utilized rotating drum technology to convert 4,800 kg of wet waste into 1,000 kg of compost, thereby generating revenue from compost sales and alleviating landfill pressure. Municipal pilots in Thrissur, Kerala utilize the Sakthan Thampuran Bio-Waste Treatment Plant to process approximately four tons of vegetable market waste daily, generating manure through organic waste converters. Research published in peer-reviewed journals indicates that compost generated under regulated conditions in Indian urban areas complies with quality standards, exhibiting safe levels of heavy metals and sufficient nutrient content, thereby rendering it suitable for agricultural applications (Jain et al., 2013). Challenges persist, such as restricted composting space in urban areas, odor control, and competition from synthetic fertilizers (Nanda & Berruti, 2021; Hammed & Sridhar, 2020). Evidence thus highlights composting's scalability and socioeconomic potential, particularly when reinforced by segregation policies, awareness campaigns, and farmer linkages.

2.8.2 Ethiopia and South Africa—Anaerobic Digestion Linking Waste, Energy, and Fertility

Pilot AD projects in Ethiopia and South Africa illustrate the benefits of decreasing organic waste volumes while simultaneously producing renewable energy. In Ethiopia, small-scale digesters supported by donors at the community level have supplied biogas for cooking and digestate for fertilizer use, thereby improving energy security and soil fertility (Godfrey et al., 2020). In South Africa, municipal biogas plants utilize market waste to generate energy for adjacent facilities, demonstrating the capacity of AD to integrate waste management, energy production, and agricultural practices. Comparative life cycle assessments (LCAs) of composting and AD in Indian cities indicate that AD possesses a greater potential for greenhouse gas mitigation than landfilling, particularly when waste is source-segregated (Ouedraogo, 2024). Scaling AD in developing economies is hindered by elevated capital costs, a shortage of skilled technicians, and variability in feedstock (Nanda & Berruti, 2021). Data from African pilots indicates that AD systems are successful only when reliable segregated organics, adequate financing, and technical training are ensured. The findings endorse multi-criteria frameworks for assessing the feasibility of AD based on economic, technical, and institutional conditions prior to adoption (Farooq et al., 2021).

2.8.3 India—Waste-to-Energy under Mixed, Moist Waste Conditions

India serves as a significant case study illustrating the constraints of WtE technologies in developing economies. Numerous urban WtE facilities have experienced suboptimal performance attributed to unsorted, moisture-laden waste streams primarily composed of organic materials, leading to diminished calorific value and heightened emissions management expenses (Nanda & Berruti, 2021). Mechanical–biological treatment (MBT) plants, exemplified by the 100 TPD facility in Goa, aim to stabilize mixed municipal waste via biological drying and mechanical sorting prior to incineration; however, their effectiveness has been constrained (Tyagi et al., 2021). Public opposition related to emissions, health impacts, and the belief that WtE undermines recycling has impeded its adoption (Ganguly & Chakraborty, 2021). WtE projects can achieve success under conditions of presorted waste, utilization of refuse-derived fuel (RDF) in cement kilns, and the enforcement of emissions standards by regulatory frameworks. Comparative assessments indicate that WtE necessitates effective governance, elevated segregation rates, and substantial financing—criteria presently observed in developed economies like Germany (Azevedo et al., 2021). The Indian experience illustrates the potential of WtE systems while also emphasizing the associated risks of financial unsustainability and environmental controversy when implemented prematurely.

2.8.4 Johannesburg, South Africa—Hybrid Recycling with MRFs and Informal Integration

Johannesburg illustrates the significance of hybrid recycling systems that integrate formal infrastructure with informal participants. Informal waste pickers play a crucial role in the recovery of plastics and recyclables, although they do so under precarious conditions. MRFs supported by municipalities and NGOs have sought to formalize recycling; however, their sustainability is compromised by inconsistent financing and inadequate governance (Kubanza & Simatele, 2020). Research conducted in South Africa indicates that contamination often limits plastics recovery, thereby diminishing recycling yields (Browning et al., 2021). Comparable insights arise from various African cities, where deficiencies in institutional planning and funding hinder formal recycling efforts (Godfrey et al., 2020). Insights from Germany's dual recycling system, which relies on household segregation, extended producer responsibility, and performance-based contracts, illustrate how upstream design features contribute to the stabilization of recycling markets (Azevedo et al., 2021). Transferring these systems to African contexts necessitates the explicit integration of informal workers to guarantee inclusion and social protection. The findings underscore that successful recycling systems in developing economies rely on hybrid models that support informal actors rather than displacing them, alongside MRFs that are underpinned by stable policy, financial, and market frameworks.

2.8.5 Southeast Asia and Sub-Saharan Africa—Digital/Industry 4.0 Pilots for Integrated Operations

Recent pilot programs in Southeast Asia and Africa demonstrate the potential of digital technologies in waste management. Batam Island, Indonesia, has implemented Industry 4.0 tools, including IoT devices, data platforms, and smart routing, to enhance material flow optimization and bolster resilience in waste-related manufacturing operations. Reviews of Waste Management 4.0 indicate that the integration of optical sorting and AI-driven systems enhances material purity when used alongside manual sorting. Additionally, IoT bins improve collection efficiency by preventing overflows. Nevertheless, these technologies continue to be costly and necessitate infrastructure and technical expertise that numerous municipalities do not possess (Cheah et al., 2022). African experiences further demonstrate that weak governance and lack of stable financing often undermine digital projects, as seen in Johannesburg's fragmented recycling governance (Kubanza & Simatele, 2020). Insights from Germany indicate that effective integration of digital technologies necessitates financing supported by extended producer responsibility, robust standards, and the involvement of informal actors via registration and mobile payment systems (Azevedo et al., 2021). Collectively, these pilots highlight that while Industry 4.0 solutions can drive efficiency and transparency, they must be embedded within strong institutional systems to succeed in developing economies.

2.9 RECYCLING AND RECOVERY

2.9.1 Significance of Recycling in the CE

Recycling is acknowledged as a fundamental component of the CE, as it enables the closure of material loops, decreases reliance on virgin resources, and lessens environmental impacts. In developing economies, the acceleration of waste generation due to urbanization and industrialization underscores the importance of recycling, which offers significant environmental and socioeconomic advantages. Recycling diverts plastics, metals, paper, and organic waste from landfills, thereby reducing pollution, mitigating greenhouse gas emissions, and supporting climate change objectives. In addition to waste diversion, it enhances industrial competitiveness and fosters job creation, particularly by incorporating informal sector participants (Kubanza & Simatele, 2020).

Researchers assert that recycling should be regarded as an integral component of a broader systemic shift towards circularity, rather than merely an isolated end-of-pipe remedy. Mwanza (2021) emphasizes that recycling enhances resource efficiency within production–consumption systems by facilitating the reintegration of secondary materials into manufacturing processes. Rumyantseva et al. (2020) assert that recycling constitutes a technical and socioeconomic process, necessitating policies, infrastructure, and behavioral changes to transition societies towards circular models. Mwanza et al. (2018) identified that awareness, convenience, and incentives significantly impact waste separation and recovery at the household level, highlighting the critical role of grassroots engagement in enhancing recycling systems.

The recycling of plastics has gained considerable importance in discussions surrounding the circular economy. Mwanza et al. (2018) examine the strategies employed by plastic manufacturing companies to recover and recycle plastic solid waste (PSW), thereby enhancing material efficiency and minimizing costs. Recent innovations, including chemical and solvent-based recycling, are broadening the possibilities for plastics recovery by facilitating the processing of complex and multilayered plastics that are challenging to manage mechanically (Klotz et al., 2024). The advanced methods illustrate the changing function of recycling within a sustainable CE; however, their implementation in developing economies remains constrained by financial and infrastructural limitations.

Recycling food waste enhances circularity by transforming biodegradable materials into energy or compost. De Sadeleer et al. (2020) analyze various management strategies: waste prevention, energy recovery, and recycling, demonstrating that recycling food waste via composting or AD is most consistent with CE principles, although prevention is prioritized above all. This emphasizes that although recycling is essential, it should be strategically placed within waste hierarchies that prioritize reduction and reuse prior to material recovery.

Guo et al. (2023) assert that recycling, in isolation, is insufficient for achieving sustainability on a global scale. The authors contend that although recycling alleviates strain on natural resources, it should be accompanied by product redesign, sustainable consumption practices, and innovative business models to realize comprehensive CE. This aligns with the view that recycling should extend beyond waste management to encompass the reconfiguration of production and consumption systems aimed at maximizing value retention.

In conclusion, the importance of recycling within the CE is evident in its capacity to convert waste into resources, promote green markets, and strengthen socioeconomic resilience. It functions as an environmental protection mechanism and an economic facilitator, particularly when integrated with comprehensive systems of prevention, redesign, and reuse. Numerous studies indicate that the effectiveness of recycling is contingent upon supportive policy frameworks, technological advancements, and behavioral changes across all societal levels. Recycling is essential; however, it is insufficient alone to facilitate the transition to sustainable, CEs.

Figure 2.6 depicts the waste management hierarchy within the context of a CE, emphasizing the position of recycling in relation to alternative strategies. The

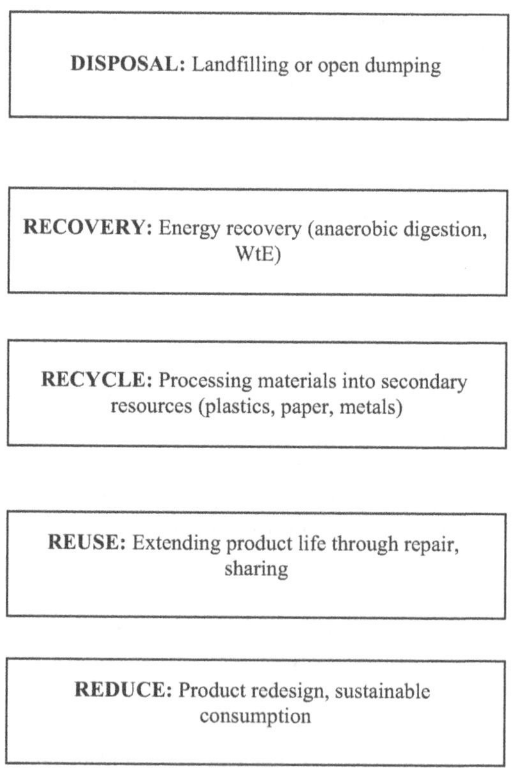

FIGURE 2.6 Recycling within the CE hierarchy.

optimal approach, Reduce, entails product redesign and sustainable consumption to decrease waste generation at the source. Reuse prolongs product lifespans via repair, sharing, and refurbishment. Recycling transforms collected materials, including plastics, metals, and organics, into secondary resources for manufacturing purposes. Recovery extracts value from residual waste via processes including AD and WtE technologies. Ultimately, disposal methods such as landfilling or open dumping are the least favorable options, as they result in the loss of material and energy value. The hierarchy underscores that recycling is essential, yet it must be complemented by upstream strategies of reduction and reuse to realize genuinely circular and sustainable systems.

2.9.2 Recycling Systems and Processes

Recycling systems are a critical component of sustainable waste management and the CE. They convert waste into valuable secondary resources, thereby minimizing the use of virgin materials and mitigating environmental impacts. Recycling systems in both developing and developed economies involve the collection, segregation, sorting, processing, and reintegration of materials into industrial supply chains. Their effectiveness relies on the efficiency of the employed technologies and the integration of both informal and formal actors, who are responsible for capturing and valorizing recyclable streams.

2.9.2.1 Plastic Recycling Processes

Plastics constitute a significant challenge while also representing a valuable waste stream. Smith et al. (2022) describe the recycling of polyethylene terephthalate (PET) as comprising multiple mechanical processes, such as sorting, washing, and pelletizing, which facilitate the reintegration of recycled PET (rPET) into packaging and textiles. Mechanical recycling is the predominant global method owing to its cost-effectiveness; however, it encounters challenges associated with contamination and downcycling (Shen & Worrell, 2024). Chemical recycling methods, including pyrolysis and solvent-based depolymerization, are increasingly recognized for their effectiveness in addressing these issues. Jeswani et al. (2021) demonstrate that the pyrolysis of mixed plastics can yield hydrocarbons suitable for chemical feedstocks; however, its environmental impacts are greater than those of mechanical recycling but less than those associated with energy recovery. Kemona and Piotrowska (2020) emphasize the recycling of polyurethanes via mechanical and chemical methods; however, widespread implementation is limited by technical and economic challenges. The studies highlight the necessity of hybrid recycling systems that integrate mechanical techniques for high-quality streams with chemical methods for complex, contaminated plastics.

2.9.2.2 Recovery of Critical Resources

Recycling extends beyond plastics. Other systems focus on the recovery of critical resources to facilitate the closure of material loops. Jupp et al. (2021) discuss phosphorus recycling from sewage sludge and industrial effluents as a crucial approach for ensuring food security and sustainability, given that phosphorus is a limited, nonrenewable resource necessary for fertilizers. Electronic waste (e-waste) recycling systems effectively recover valuable metals, including gold, silver, and rare earth elements, which would otherwise be lost in inadequately managed dumpsites. Rene et al. (2021) observe that contemporary e-waste recycling utilizes hydrometallurgical and pyrometallurgical techniques, in addition to developing bioleaching methods, to recover resources and mitigate environmental impact. In developing economies, informal recyclers frequently lead e-waste dismantling activities; however, the absence of protective equipment and regulatory oversight leads to significant health risks and environmental contamination. The integration of informal practices into formal recycling systems presents a significant challenge.

2.9.2.3 System Integration and Life Cycle Perspectives

Recycling systems should be analyzed through life cycle lens. Jeswani et al. (2021) illustrate that although chemical recycling can expand the variety of recoverable plastics, it is not necessarily more environmentally beneficial in all cases. Mechanical recycling is advantageous when clean streams are present, highlighting the significance of source segregation and collection efficiency. Incorporating chemical recycling, phosphorus recovery, and e-waste resource recovery illustrates the necessity for recycling systems to be multidimensional and adaptable. The integration of technologies, supportive policies, and market incentives ensures that recycling systems divert waste from landfills and reinforce circular flows of critical resources.

In summary, recycling systems and processes (Figure 2.7) encompass a variety of technologies and practices. The discussion encompasses plastics, e-waste, and phosphorus recovery, emphasizing recycling's potential to enhance resource efficiency, mitigate environmental impacts, and generate economic value. Their effectiveness depends on strong governance, robust infrastructure, technological innovation, and the inclusion of informal actors in developing economies.

Figure 2.7 depicts the sequential process of recycling systems, beginning with waste collection, which encompasses both formal municipal services and informal participants like waste pickers. Effective recycling relies significantly on segregation, as source separation and manual sorting facilitate cleaner streams of plastics, metals, glass, e-waste, and organics. Upon segregation, materials proceed through two primary processing pathways. Mechanical processing encompasses shredding, washing, and pelletizing, which are typically utilized for materials like PET plastics, metals, and glass, facilitating their reintegration into manufacturing processes. Chemical processing encompasses advanced technologies such as pyrolysis, depolymerization, and solvent-based methods, which are especially effective for complex plastics and contaminated waste streams.

Figure 2.7 illustrates resource recovery systems for specialized materials. This encompasses the extraction of valuable metals and rare earth elements from electronic waste, the recycling of phosphorus from wastewater and sludge for fertilizer production, and the reclamation of polyurethanes and other polymers. The final stage involves reintegration into industry, utilizing recycled materials in packaging, textiles, construction, fertilizers, and electronics, thereby completing the cycle in the circular economy.

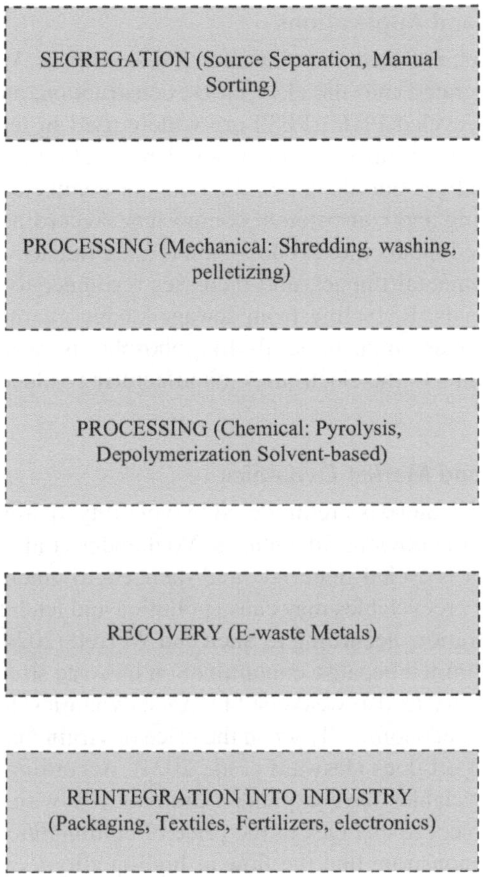

FIGURE 2.7 Recycling systems and processes in a CE.

The flow highlights that recycling constitutes a multifaceted system, incorporating various technologies and stakeholders to enhance resource efficiency, mitigate environmental impacts, and generate economic opportunities.

2.9.3 Market for Recycled Materials

The advancement of the CE and the effectiveness of recycling systems depend heavily on the growth of markets for recycled materials. Recovered materials run the risk of being disposed of, downcycled, or stored in the absence of sustainable markets, which would undermine both economic and environmental goals. By giving companies access to secondary resources, lowering dependency on virgin raw materials, and fostering innovation throughout supply chains, markets turn recycling from a cost center into a source of value creation.

2.9.3.1 Demand and Applications

In a variety of sectors, recycled materials are vital feedstocks. While recycled metals and glass are reintegrated into the electronics, construction, and automotive industries, plastics like recycled PET (rPET) are widely used in textiles and packaging (Smith et al., 2022). According to Leverenz and Kreith (2002), recycled materials not only serve as alternatives but also open the door for completely new product lines, such as eco-packaging and construction composites. According to Field and Sroufe (2007), using recycled inputs into production has a big impact on supply chain strategy, lessens environmental impact, and increases resilience to changes in the price of virgin raw materials. Recycling from sewage sludge guarantees the stability of fertilizer supplies for essential minerals like phosphorus, while recycling e-waste returns valuable metals to the electronics manufacturing industry (Jupp et al., 2021; Rene et al., 2021).

2.9.3.2 Barriers and Market Dynamics

Markets for recycled materials are hindered by volatility, quality issues, and irregular demand, despite the obvious advantages. Vogtländer et al. (2017) stress that the commercial value of recycled materials and their environmental impact are intertwined; lower-quality recyclables may cause pollution and fetch lower prices. Plastics offer a glaring illustration: according to Shen and Worrell (2024), rPET's use in food grade packaging is limited because contamination in waste streams compromises its quality. Market stability is also weakened by price dynamics. Recycled plastics find it difficult to compete economically when the price of virgin materials declines, such as when the price of oil does (Jeswani et al., 2021). According to Sadowski (2010), EU markets for recyclables have expanded dramatically with the help of policies, but they are still subject to changes in the price of commodities globally. Similarly, Mo et al. (2009) demonstrate that the flow of high-quality recyclables into markets is limited in Suzhou, China, due to fragmented collection and poor coordination in the recycling systems.

2.9.3.3 Policy, Regulation, and Market Creation

Strong policy frameworks play a decisive role in shaping recycled material markets. EPR, recycled content mandates, and procurement policies create stable demand by obligating industries to use secondary resources. Sadowski (2010) emphasizes that EU regulations have stimulated market growth by embedding recycling targets and quality standards into policy. In the US, growing corporate commitments to use recycled inputs in packaging have expanded demand for rPET and recycled aluminum (Smith et al., 2022). Mo et al. (2009) illustrate how China has advanced recycling policies by formalizing collection systems and integrating informal recyclers, though challenges in enforcement and coordination persist.

2.9.3.4 Market Innovation and Supply Chains

Markets for recycled materials are changing because of new developments. According to Field and Sroufe (2007), supply chain strategies need to change to include long-term contracts for recycled inputs, quality assurance, and traceability.

This illustrates how recycling is becoming acknowledged as a crucial operational tactic in sustainable manufacturing as well as a waste management activity. As online platforms increasingly link recyclers with industrial purchasers, lowering transaction costs and boosting transparency, digital technologies further improve market efficiency (Cheah et al., 2022).

2.10 DISPOSAL AND LANDFILLING

2.10.1 Traditional Disposal Methods and Their Demerits

Open dumping, open burning, and burying waste are all examples of traditional disposal techniques used in poor economies; each has drawbacks and disadvantages of its own.

2.10.1.1 Open Dumping

The most popular technique in many poor nations is open dumping, which is disposing of waste in uncontrolled locations without protective measures like gas management systems, leachate collection, or liners. Although it seems low cost in the short run, it has serious long-term effects. According to Lema et al. (1988), organic contaminants, heavy metals, and xenobiotic chemicals are frequently found in leachate produced from decaying waste. These substances contaminate surrounding rivers and lakes and seep into groundwater. Anaerobic decomposition produces methane, which is a major contributor to greenhouse gas emissions and exacerbates climate change. Socially, the health and quality of life of populations residing close to open dumps are negatively impacted by offensive odors, pest infestations, and exposure to dangerous materials. Economically speaking, open dumping lowers land prices and raises remediation expenses for municipalities when sites turn dangerous. Critical recovery services are provided by informal waste pickers who remove recyclables from these dumps, but they operate in hazardous environments where they run the risk of becoming hurt, infected, or exposed to toxins (Grossule & Stegmann, 2020). Therefore, the apparent financial savings of open dumping conceal more serious harm to the environment, health issues, and financial losses.

2.10.1.2 Open Burning

In situations where collection and disposal services are insufficient, open burning of waste is a typical practice to minimize volume. Even while it swiftly removes obvious garbage mounds, its effects are quite detrimental. Dioxins, furans, and particulate matter are released when plastics and mixed municipal debris are burned. These substances have been linked to cancer, cardiovascular disease, and respiratory ailments (Ghasemi & Yusuff, 2016). Beyond dumpsites, smoke and airborne pollutants spread, reducing the quality of the air in cities and endangering entire communities. Due to their preexisting vulnerability and lack of access to healthcare, those living in informal settlements close to burning areas suffer disproportionately on a social level. Economically speaking, the small short-term gain of volume reduction is outweighed by the expenses of treating illnesses brought on by pollution and lost productivity. Additionally, burning waste destroys recyclables that may be used to make secondary materials or energy, like paper, plastics, and organics (Narayana, 2009). In addition to harming the environment, this is a lost chance to boost the recycling and job creation sectors.

2.10.2 Modern Landfilling Techniques

It is commonly acknowledged that modern landfilling is a crucial part of integrated solid waste management. It includes environmental monitoring and engineered safeguards, unlike uncontrolled dumping. But despite progress, there are still issues, especially in underdeveloped nations.

2.10.2.1 Engineering and Design Features

Leachate collecting systems, gas wells, and designed liners are features of contemporary sanitary landfill construction. While collecting systems direct liquid to treatment facilities, clay or high-density polyethylene (HDPE) liners stop harmful leachate from leaking into soil (Sharma et al., 2021). Multilayer capping prolongs landfill life, while daily cover materials reduce odor, vermin, and wind-blown litter (Vijayalakshmi, 2020). Sanitary landfills are safer and more palatable than conventional dumps because of these characteristics.

Restrictions and prospects for the future: Cost is the primary barrier; capping, monitoring systems, and designing liners all demand large financial outlays. Sanitary landfills and uncontrolled dumping coexist in many developing nations because of governance and financial constraints (Narayana, 2009). To make designed designs economical and scalable, future paths will require implementing real-time monitoring technologies and cost-effective liners.

2.10.2.2 Leachate Quality and Treatment

One of the biggest problems is leachate, the liquid that drains from waste. While older leachates contain refractory chemicals that are resistant to biodegradation, younger leachates are rich in organic matter and ammonium (Modin, 2012). Treatment methods include sophisticated procedures like oxidation and reverse osmosis, as well as biological systems like wetlands and sequencing batch reactors (Sharma et al., 2021). These technologies lessen the possibility of surface and groundwater contamination.

Restrictions and prospects for the future: Many towns release untreated leachate because they lack the technical competence to operate sophisticated treatment systems (Schiopu & Gavrilescu, 2010). Sludge production and exorbitant energy expenses are still issues. Low-energy technologies designed for emerging economies and hybrid treatment systems that combine biological and physicochemical techniques are examples of future directions.

2.10.2.2.1 Microbiology and Biogas Capture

Dynamic microbial ecosystems seen in landfills promote methane production and breakdown. A pollutant can be converted into a renewable energy source by using gas wells to capture methane (Nanda & Berruti, 2021). Meyer-Dombard et al. (2020) emphasize that increasing stabilization rates and gas production requires an understanding of landfill microorganisms. Utilizing biogas minimizes greenhouse gas emissions, produces power, and lessens dependency on fossil fuels.

Restrictions and prospects for the future: Significant infrastructure is needed for gas recovery systems, and badly maintained sites permit fugitive methane emissions, which undermines climate goals. Furthermore, performance is variable due to the incomplete understanding of microbial communities in landfills. To maximize gas production, future studies should concentrate on "next-generation" microbiology, which includes real-time microbial monitoring and tailored microbial consortia.

2.10.2.2.2 Bioreactor Landfills

By adding moisture, recirculating leachate, and occasionally aerating cells, bioreactor landfills purposefully increase microbial activity. This shortens post-closure monitoring, increases methane outputs, and speeds up waste stabilization (Sharma et al., 2021). By combining energy recovery and disposal, they are more in line with sustainability objectives. According to Vijayalakshmi (2020), bioreactor designs can contribute even more to circular resource flows when combined with renewable energy systems.

Restrictions and prospects for the future: Despite their potential, bioreactor landfills are not widely adopted in environments with limited resources because of their high capital costs and technical complexity. Operational difficulties still include controlling the moisture balance and avoiding leachate spills. Creating modular bioreactor designs for gradual improvements and combining them with digital monitoring systems for peak performance are examples of future approaches.

2.10.3 Health and Environmental Effects of Illegal Solid Waste Disposals

There are major health and environmental hazards associated with the illegal disposal of solid waste, whether it is through fly-tipping, uncontrolled dumping, or open burning. Toxic air pollutants, damaged soils, and polluted rivers are all hazards to communities close to unlawful dumpsites. Unmanaged waste releases methane, volatile organic compounds, and leachates rich in ammonia and heavy metals, endangering ecosystems and human health, according to Nanda and Berruti (2021). Similarly, untreated leachates from unregulated sites can contaminate aquifers, increasing the risk of waterborne illnesses such as cholera and dysentery (Lema et al., 1988).

According to Ghasemi and Yusuff (2016), the open burning of plastics, mixed municipal garbage, and medical waste results in particulate matter and carcinogenic substances that worsen respiratory conditions and raise the chance of developing cancer in the long run. Children and other informal garbage workers are disproportionately impacted because, while collecting and sorting, they are in close proximity to sharp objects, infectious debris, and dangerous chemicals. India's dependence on uncontrolled dumpsites has resulted in serious urban health issues, particularly in the peri-urban slums that are close to these sites, according to Narayana (2009).

Illegal dumping damages aquatic habitats, disturbs soil quality, and endangers species from an ecological standpoint. According to Grossule and Stegmann (2020), unlawful dumping feeds unsustainable cycles of pollution and resource depletion in the absence of strict monitoring and alternatives. Enforcing waste management laws, funding official collection systems, and implementing inclusive solutions that incorporate unofficial actors into safer, regulated frameworks are all necessary to address these problems.

REFERENCES

Abubakar, I.R., Maniruzzaman, K.M., Dano, U.L., AlShihri, F.S., AlShammari, M.S., Ahmed, S.M.S., Al-Gehlani, W.A.G. and Alrawaf, T.I. (2022) 'Environmental sustainability impacts of solid waste management practices in the global South', *International Journal of Environmental Research and Public Health*, 19(19), p. 12717.

Alzamora, B.R. and Barros, R.T.V. (2020). 'Review of municipal waste management charging methods in different countries', *Waste Management*, 115, pp. 47–55. https://doi.org/10.1016/j.wasman.2020.07.020.

Azam, M., Jahromy, S.S., Raza, W., Raza, N., Lee, S.S., Kim, K.H. and Winter, F. (2020) 'Status, characterization, and potential utilization of municipal solid waste as a renewable energy source: Lahore case study in Pakistan', *Environment International*, 134, p. 105291.

Azevedo, B.D., Scavarda, L.F., Caiado, R.G.G. and Fuss, M. (2021) 'Improving urban household solid waste management in developing countries based on the German experience', *Waste Management*, 120, pp. 772–783.

Browning, S., Beymer-Farris, B. and Seay, J.R. (2021) 'Addressing the challenges associated with plastic waste disposal and management in developing countries', *Current Opinion in Chemical Engineering*, 32, p. 100682.

Bryan, G., Glaeser, E. and Tsivanidis, N. (2020) 'Cities in the developing world', *Annual Review of Economics*, 12(1), pp. 273–297.

Bui, T.D., Tseng, J.W., Tseng, M.L. and Lim, M.K. (2022) 'Opportunities and challenges for solid waste reuse and recycling in emerging economies: A hybrid analysis', *Resources, Conservation and Recycling*, 177, p. 105968.

Chandrappa, R. and Das, D.B. (2024) *Solid Waste Management: Principles and Practice*. Cham: Springer Nature.

Cheah, C.G., Chia, W.Y., Lai, S.F., Chew, K.W., Chia, S.R. and Show, P.L. (2022) 'Innovation designs of Industry 4.0 based solid waste management: Machinery and digital circular economy', *Environmental Research*, 213, p. 113619.

Chen, D.M.C., Bodirsky, B.L., Krueger, T., Mishra, A. and Popp, A. (2020) 'The world's growing municipal solid waste: Trends and impacts', *Environmental Research Letters*, 15(7), p. 074021.

De Sadeleer, I., Brattebø, H. and Callewaert, P. (2020). 'Waste prevention, energy recovery or recycling? Directions for household food waste management in light of circular economy policy', *Resources, Conservation and Recycling*, 160, 104908. https://doi.org/10.1016/j.resconrec.2020.104908

Debrah, J.K., Vidal, D.G. and Dinis, M.A.P. (2021) 'Raising awareness on solid waste management through formal education for sustainability: Evidence from developing countries', *Recycling*, 6(1), p. 6.

Farooq, A., Haputta, P., Silalertruksa, T. and Gheewala, S.H. (2021) 'A framework for the selection of suitable waste-to-energy technologies for a sustainable municipal solid waste management system', *Frontiers in Sustainability*, 2, p. 681690.

Field, J.M. and Sroufe, R.P. (2007). 'The use of recycled materials in manufacturing: implications for supply chain management and operations strategy', *International Journal of Production Research*, 45(18–19), pp. 4439–4463.

Ganguly, R.K. and Chakraborty, S.K. (2021) 'Integrated approach to municipal solid waste management during COVID-19: A developing country perspective', *Case Studies in Chemical and Environmental Engineering*, 3, p. 100087.

Ghasemi, M.K. and Yusuff, R.B.M. (2016) 'Advantages and disadvantages of healthcare waste treatment and disposal alternatives: Malaysian scenario', *Polish Journal of Environmental Studies*, 25(1), pp. 17–25.

Godfrey, L., Ahmed, M.T., Gebremedhin, K.G., Katima, J.H., Oelofse, S., Osibanjo, O. and Richter, U.H. (2020). Solid waste management in Africa: Governance failure or development opportunity?, in I. Kararach, J. Odhiambo and H. Nyamongo (eds.) *Regional Development in Africa.* Abingdon: Routledge, pp. 235–261.

Grossule, V. and Stegmann, R. (2020) 'Problems in traditional landfilling and proposals for solutions based on sustainability', *Detritus*, 12, pp. 78–91.

Guo, J., Ali, S. and Xu, M. (2023). 'Recycling is not enough to make the world a greener place: Prospects for the circular economy', *Green Carbon*, 1(2), pp. 123–136. https://doi.org/10.1016/j.greenca.2023.10.006

Hammed, T.B. and Sridhar, M.K.C. (2021). Green technology approaches to solid waste management in the developing economies, in Nicolas Oguge, Daniel Olago, Lianet Peña del Valle and Douglas Ayal (eds.) *African Handbook of Climate Change Adaptation*. Cham: Springer, pp. 1293–1312. https://doi.org/10.1007/978-3-030-45106-6_174.

Harfadli, M.M.A., Ramadan, B.S., Rachman, I. and Matsumoto, T. (2024) 'Challenges and characteristics of the informal waste sector in developing countries: An overview', *Journal of Material Cycles and Waste Management*, 26(3), pp. 1294–1309.

Jagun, Z.T., Daud, D., Ajayi, O.M., Samsudin, S., Jubril, A.J. and Rahman, M.S.A. (2023) 'Waste management practices in developing countries: A socio-economic perspective', *Environmental Science and Pollution Research*, 30(55), pp. 116644–116655.

Jain, M.S., Daga, M. and Kalamdhad, A.S. (2013). 'Characterisation of municipal solid waste compost from selected Indian cities', *Journal of Environmental Protection*, 4(2), pp. 90–98. https://doi.org/10.4236/jep.2013.42011.

Jeswani, H., Krüger, C., Russ, M., Horlacher, M., Antony, F., Hann, S. and Azapagic, A. (2021) 'Life cycle environmental impacts of chemical recycling via pyrolysis of mixed plastic waste', *Science of the Total Environment*, 769, p. 144483.

Jupp, A.R., Beijer, S., Narain, G.C., Schipper, W. and Slootweg, J.C. (2021) 'Phosphorus recovery and recycling: Closing the loop', *Chemical Society Reviews*, 50(1), pp. 87–101.

Khan, I. and Kabir, Z. (2020). Waste-to-energy generation technologies and the developing economies: A multi-criteria analysis for sustainability assessment', *Renewable Energy*, 150, pp. 320–333. https://doi.org/10.1016/j.renene.2019.12.132.

Klotz, M., Oberschelp, C., Salah, C., Subal, L. and Hellweg, S. (2024). 'The role of chemical and solvent-based recycling within a sustainable circular economy for plastics', *Science of the Total Environment*, 906, 167586. https://doi.org/10.1016/j.scitotenv.2023.167586.

Kubanza, N.S. and Simatele, M.D. (2020) 'Sustainable solid waste management in developing countries: Institutional strengthening in Johannesburg', *Journal of Environmental Planning and Management*, 63(2), pp. 175–188.

Kumari, T. and Raghubanshi, A.S. (2023). Waste management practices in the developing nations: Challenges and opportunities, in P. Agamuthu and A. Maalouf (eds.) *Waste Management and Resource Recycling in the Developing World.* Singapore: Springer, pp. 773–797.

Lema, J.M., Mendez, R. and Blazquez, R. (1988) 'Characteristics of landfill leachates and alternatives for their treatment: A review', *Water, Air, and Soil Pollution*, 40(3), pp. 223–250.

Leverenz, H. and Kreith, F. (2002). 'Markets and products for recycled material', *Handbook of Solid Waste Management*, 2, pp. 1–9.

Lusaka City Council (2021). *Solid Waste Management Annual Report 2021.* Lusaka: Lusaka City Council.

Meyer-Dombard, D.A.R., Bogner, J.E. and Malas, J. (2020) 'A review of landfill microbiology and ecology', *Frontiers in Microbiology*, 11, p. 1127.

Mo, H., Wen, Z. and Chen, J. (2009) 'China's recyclable resources recycling system and policy: A case study in Suzhou', *Resources, Conservation and Recycling*, 53(7), pp. 409–419. Available at: https://doi.org/10.1016/j.resconrec.2009.03.002.

Modin, H. (2012). *Modern Landfill Leachates: Quality and Treatment.* Doctoral thesis. Lund: Department of Water Resources Engineering, Lund University.

Mwanza, B.G. (2021). Introduction to recycling. In *Recent developments in plastic recycling* (pp. 1–13). Singapore: Springer Singapore.

Mwanza, B.G., Mbohwa, C. and Telukdarie, A. (2017) 'Strategies for incorporating informal waste collectors into formalised systems', in *Proceedings of the International Conference on Industrial Engineering and Operations Management*. Bogota, Colombia.

Mwanza, B.G., Mbohwa, C. and Telukdarie, A. (2018). 'Municipal solid waste management in Kitwe City: An engineering management perspective', *Management of Environmental Quality: An International Journal.*

Nanda, S. and Berruti, F. (2021) 'Municipal solid waste management and landfilling technologies: A review', *Environmental Chemistry Letters*, 19(2), pp. 1433–1456.

Narayana, T. (2009) 'Municipal solid waste management in India: From waste disposal to resource recovery', *Waste Management*, 29(3), pp. 1163–1166.

Ouedraogo, A.S. (2024). 'Comparative life cycle assessment of landfilling with integrated municipal solid waste management systems', *Environments*, 11(11), 248. https://doi.org/10.3390/environments11110248.

Rene, E.R., Sethurajan, M., Ponnusamy, V.K., Kumar, G., Dung, T.N.B., Brindhadevi, K. and Pugazhendhi, A. (2021) 'Electronic waste generation, recycling and resource recovery', *Journal of Hazardous Materials*, 416, p. 125664.

Rumyantseva, A., Rumyantseva, E., Berezyuk, M. and Plastinina, J. (2020) 'Waste recycling as an aspect of the transition to a circular economy', *IOP Conference Series: Earth and Environmental Science*, 534(1), Article 012002. https://doi.org/10.1088/1755-1315/534/1/012002

Sadowski, A. (2010). *'The development prospects for the recycled materials markets in the European Union', Acta Universitatis Lodziensis. Folia Oeconomica*, 241, pp. 73–88.

Satori, M., Prasetyo, L.B., Suyanto, S. and others, (2020). *The Role of the Informal Sector in Waste Management: A Case Study of Bandung, Indonesia.* Bandung: Institut Teknologi Bandung.

Schiopu, A.-M. and Gavrilescu, M. (2010). 'Options for the treatment and management of municipal landfill leachate: Common and specific issues', *Clean – Soil, Air, Water*, 38(12), pp. 1101–1110. https://doi.org/10.1002/clen.201000068.

Sengupta, D., Ilankoon, I.M.S.K., Kang, K.D. and Chong, M.N. (2022) 'Circular economy and household e-waste management in India', *Minerals Engineering*, 184, p. 107661.

Sharma, A., Ganguly, R. and Gupta, A.K. (2021). 'Impact of sanitary landfill design and leachate management on environmental sustainability: A review', *Environmental Technology & Innovation*, 21, 101315. https://doi.org/10.1016/j.eti.2020.101315.

Sharma, K.D. and Jain, S. (2020) 'Municipal solid waste generation, composition and management: Global scenario', *Social Responsibility Journal*, 16(6), pp. 917–948.

Shen, L. and Worrell, E. (2024) 'Plastic recycling', in Worrell, E. and Reuter, M.A. (eds.) *Handbook of Recycling*. Amsterdam: Elsevier, pp. 497–510.

Smith, R.L., Takkellapati, S. and Riegerix, R.C. (2022) 'Recycling of plastics in the United States', *ACS Sustainable Chemistry & Engineering*, 10(6), pp. 2084–2096.

Tong, Y.D., Huynh, T.D.X. and Khong, T.D. (2021) 'Role of the informal sector in sustainable municipal solid waste management', *Waste Management*, 124, pp. 118–127.

Tyagi, V.K., Kapoor, A., Arora, P., Banu, J.R., Kazmi, A.A. and others, (2021). 'Mechanical-biological treatment of municipal solid waste: Case study of 100 TPD Goa plant, India', *Journal of Environmental Management*, 292, 112741. https://doi.org/10.1016/j.jenvman.2021.112741

Ugwu, S.N., Chukwuma, C.S., Nwakaire, J.N., Aghoghovwia, O.A., Okechukwu, M.E. and Babatunde, D.E. (2020) 'Sustainable waste management and circular economy in developing countries: A review', *Environmental Challenges*, 2, Article 100024. Available at: https://doi.org/10.1016/j.envc.2020.100024.

Vaverková, M.D. et al. (2021) 'Municipal solid waste management under COVID-19', *Environmental Geotechnics*, 8(3), pp. 217–232.

Vijayalakshmi, M. (2020). 'Modern waste management techniques: A critical review', *International Journal of Creative Research Thoughts (IJCRT)*, 8(10), pp. 3277–3287.

Vogtlander, J.G., Scheepens, A.E., Bocken, N.M. and Peck, D. (2017). 'Combined analyses of costs, market value and eco-costs in circular business models: eco-efficient value creation in remanufacturing', *Journal of Remanufacturing*, 7(1), pp. 1–17.

Zambia Environmental Management Agency (ZEMA) (2020) *State of the Environment Report: Solid Waste Management in Zambia*. Lusaka: ZEMA.

3 Implementing Circular Economy in Solid Waste Management

3.1 POLICY AND REGULATORY FRAMEWORKS

3.1.1 Existing Regulations and Policies in Developing Economies

A range of legislative and regulatory frameworks that aim to enhance waste recovery, lessen reliance on landfills, and conform to the principles of the CE influence solid waste management (SWM) in developing nations. The landscape is dominated by four primary regulatory instrument types: national waste management policies, MSW ordinances or bylaws, plastic bag bans and levies, and EPR. Even though these frameworks are essential first steps toward sustainable waste management, their efficacy has been variable due to institutional fragmentation, poor enforcement, and a lack of stakeholder integration.

3.1.1.1 Extended Producer Responsibility

Under the EPR policy approach, manufacturers are required by law to assume responsibility for the proper disposal, recycling, or collection of their products and packaging following customer use. The idea is to encourage eco-design, recyclability, and circularity by shifting the operational and financial responsibility of waste management from municipalities to producers (Mwanza & Telukdarie, 2024). Under the National Environmental Management: Waste Act (2008), South Africa implemented Section 18 EPR Regulations in 2021, mandating that manufacturers of lighting, packaging, and electronic waste register with the government, fulfill recovery goals, and contribute to Producer Responsibility Organizations. Kenya complied with the Sustainable Waste Management Act (2022) and the EPR Regulations (2024–2025), which required producers to supply recovery statistics and finance collection systems. Similar limitations were placed on packaging and plastics by Zambia's Environmental Management (EPR) Regulations, Statutory Instrument No. 65 of 2018, which also prohibited thin plastic bags smaller than 30 microns unless approved.

The application of these frameworks has been inconsistent. In their study of e-waste in China and Thailand, Kojima et al. (2009) demonstrated how the efficiency of EPR is undermined by inadequate regulation, a lack of infrastructure, and the prevalence of informal recycling. This conclusion is similar to the situation in Africa, where informal waste pickers continue to play a crucial role in recovery despite being mainly left out of official EPR programs. Leclerc and Badami (2020) point out that

DOI: 10.1201/9781003591764-3

poor consumer knowledge, high compliance costs, and institutional fragmentation all work against EPR in emerging environments. According to Peng et al. (2019), corporate environmental policies also play a role in the effectiveness of EPR; companies that include compliance with larger sustainability objectives perform better than those that see EPR as merely a regulatory burden. This is supported by recent African contributions: Mwanza and Telukdarie (2024) contend that EPR is a crucial hub for CE transitions but necessitates transparent data systems, supportive policy ecosystems, and SMEs' inclusion. The weakest link is enforcement capacity, according to Mwanza and Mbohwa (2019); in the absence of robust monitoring mechanisms and sanctions, EPR runs the risk of being symbolic rather than revolutionary.

3.1.1.2 Plastic Bag Bans and Levies

One of Africa's most noticeable waste policies is the prohibition and taxation of plastic bags. To lessen litter and pollution in the environment, these policies forbid or discourage the use of single-use plastic carrier bags. In 2017, Kenya declared a statewide ban with severe punishments for violators. This strategy was first implemented in Rwanda in 2008 when polythene bags were outlawed. Law No. 17/2019 later expanded the ban to include single-use plastics. In 2019, Tanzania implemented a similar statewide ban, while in 2004, South Africa imposed minimum thickness requirements and a plastic bag charge. Under the 2018 EPR legislation, Zambia additionally limits the production and distribution of thin plastic bags. Although visible waste has decreased as a result of these measures, enforcement issues still exist. Illegal manufacturing and smuggling continue, particularly in areas with a shortage of reasonably priced alternatives. Kojima et al. (2009) warn that such prohibitions frequently push plastics into black markets rather than outlawing them if they are not strictly enforced and have no workable alternatives.

3.1.1.3 Municipal Solid Waste Ordinances and Bylaws

Local laws known as municipal ordinances govern the collection, disposal, and storage of waste in cities and districts. They control tariffs, fines, and waste service provider licensing. The SWM Act of 2015 in Nairobi requires contractor license and source segregation. Landfill operators, garbage generators, and carriers must abide by Johannesburg's garbage Management Bylaw (2013, modified 2021). Franchised collection zones and hauler licensing were established by Lusaka's MSW Bylaws (2004), and they are still essential to local waste management in Zambia. However, these ordinances frequently lack sufficient funding and enforcement. According to Mwanza and Mbohwa (2019), compliance is weakened by institutional fragmentation and small municipal budgets. Additionally, informal waste pickers, who control collection and recycling, are frequently left out of formal contracts. Kojima et al. (2009) have identified this issue in Asian e-waste systems.

3.1.1.4 National Waste Management Strategies and Acts

Framework laws and strategies set long-term objectives, norms, and institutional duties for waste management at the national level. Diversion and CE goals are outlined in South Africa's National Environmental Management: Waste Act (2008) and National Waste Management Strategy (2020). Source segregation, county compliance, and EPR integration are prioritized under Kenya's Sustainable Waste Management Act (2022) and National Solid Waste Management Strategy (2014). Building on its 2004 strategy, Zambia's Solid Waste Regulation and Management Act (2018a) established a framework for service and facility licensing. Despite having lofty objectives, these tools are still not widely used. Institutional fragmentation weakens national waste governance (Leclerc & Badami, 2020), although Mwanza and Telukdarie (2024) emphasize that to guarantee systemic impact, such initiatives must be in line with EPR and CE concepts.

To improve SWM, developing economies have collectively created extensive legislative frameworks, including national plans, municipal ordinances, plastic bag bans and levies, and EPR schemes. However, their effectiveness is restricted due to a lack of funding, a lack of effective enforcement, poor data systems, and a failure to adequately integrate informal actors. Mwanza and Mbohwa (2019) and Kojima et al. (2009) show that regulations are useless without strong institutional backing and monitoring. Peng et al. (2018) remind us that private sector involvement is also important: although some companies embrace change, others reject it. Companies that align compliance with sustainability plans gain a competitive edge. Mwanza and Telukdarie (2024) place EPR at the core of CE transformations, while Leclerc and Badami (2020) stress the necessity of institutional coordination. These observations highlight the fact that although policies are in place in theory, closing the gaps in funding, enforcement, and stakeholder participation is necessary to achieve significant results in SWM in developing nations.

3.1.2 Best Practices from Developed Economies

3.1.2.1 Extended Producer Responsibility

One of the most popular best practices for promoting the CE in industrialized economies is EPR. Strong EPR frameworks in the EU hold manufacturers physically and financially accountable for overseeing the end of life of their goods. For example, laws pertaining to packaging, electronics, batteries, and automobiles require manufacturers to adhere to stringent collection and recycling goals. For instance, by 2022, businesses must guarantee that at least 63% of plastic packaging is recycled, according to Germany's Packaging Act.

Scholars emphasize that EPR is a catalyst for sustainable product innovation and goes beyond waste management. Stakeholder perspectives demonstrate how EPR can transform value chains by motivating companies to develop goods with reuse, recycling, and recovery in mind (Kunz et al., 2018). In a similar vein, Maitre-Ekern (2021) makes the case for shifting the emphasis from the conventional "end-of-pipe" to pre-market producer responsibility, wherein producers are required to incorporate sustainability and circularity considerations into the design and production stages of their products. This change guarantees that resource efficiency is ingrained throughout a product's life cycle, not only upon disposal.

In Canada, provincial waste rules have also been successfully harmonized through the advancement of EPR. According to McKerlie et al. (2006), companies were incentivized to invest in recycling infrastructure and enhance material recovery rates through provincial EPR programs for packaging, electronics, and household hazardous waste. By encouraging material loops and lowering dependency on landfills, these initiatives support the larger objectives of the CE.

All things considered, industrialized economies show that EPR can be a potent market-based and regulatory tool. When effectively implemented, it encourages eco-design, encourages recycling system innovation, and transfers the cost of waste from communities and governments to producers.

3.1.2.2 Circular Industrial Symbiosis

A key component of the CE is IS, which allows businesses to work together by recycling, reusing, or trading waste streams, energy, and resources across industries. Denmark's Kalundborg Eco-Industrial Park is the most well-known example, where businesses trade surplus steam, cooling water, gypsum, and other by-products. Some firms utilize materials that would otherwise be dumped, while others catch excess heat from a power plant to support fish farms and offer district heating.

Studies demonstrate how IS has a wider impact on sustainability. According to Cecchin et al. (2020), IS connects the CE to the discussion of sustainable development by concurrently addressing social, economic, and environmental objectives. IS promotes local employment, lowers greenhouse gas emissions, and increases resilience in addition to straightforward resource-sharing.

Baldassarre et al. (2019) contend that from a design standpoint, successful eco-industrial clusters need to incorporate knowledge from both industrial ecology (IE) and CE. Their study describes industrial park design procedures that promote creativity, trust, and sustained cooperation among businesses.

Mapping efforts demonstrate the size and diversity of IS networks throughout Europe. According to Domenech et al. (2019), typologies range from more expansive regional platforms aided by digital technology to more confined clusters like Kalundborg. Although the effectiveness of these networks varies, taken as a whole, they show a substantial contribution to resource efficiency and CE goals.

Furthermore, by guaranteeing that materials circulate through closed loops rather than escaping into waste streams, symbiotic networks serve as practical enablers of policy aspirations, as noted by Castellet-Viciano et al. (2022), who emphasize IS as a technique to assure CE implementation.

Overall, IS boosts competitiveness by encouraging collaborative innovation and lowering waste and costs, according to best practices in developed economies. Adopting IS concepts in industrial zones, like mining and manufacturing clusters, could open synergies that support climate resilience, local development, and resource efficiency in Zambia and comparable contexts.

3.1.2.3 Circular Product Design and Innovation

To prolong product life cycles, reduce waste, and guarantee that resources stay in productive use, circular product design and innovation are essential. The integration of these concepts into supply chains and production has been spearheaded by developed economies. Manufacturers are required by Japan's Home Appliance Recycling Law to design devices that are simple to disassemble and recover valuable elements from. Similar to this, businesses like Fairphone in the Netherlands manufacture modular cellphones that can be upgraded and repaired, lowering electronic waste and encouraging customers to engage in circular consumption.

The importance of product design in promoting the CE is shown by research. To assist businesses, align with the ideas of the CE, Den Hollander et al. (2017) offer a typology of design techniques, including design for lifetime, modularity, repairability, and recyclability. Expanding on this, Burke et al. (2023) point out that combining supply chain management and product design promotes circularity throughout the value chain, guaranteeing not only product recyclability but also effective material recovery and reverse logistics.

One strategic enabler of CE has been identified as new product development (NPD). According to Pinheiro et al. (2019), NPD techniques enable businesses to incorporate sustainability into the innovation phase, allowing for the creation of goods that prioritize resource efficiency and reuse. Furthermore, through a thorough analysis of the literature, Suchek et al. (2021) highlights that innovation encompasses more than just redesigning products; it also extends to business models, services, and ecosystems, enabling companies to generate profit while encouraging environmental stewardship.

These global examples show that innovation and circular design are crucial for sustainability and competitiveness, not optional. By incorporating circular design into manufacturing, Zambia can lessen its reliance on imports, encourage domestic entrepreneurship, and match industrial production with the nation's green growth objective. Examples of this include remanufacturable machinery, modular furniture, and recyclable packaging.

3.1.2.4 Deposit Return Schemes

Deposit Return Schemes (DRS), which are extensively used in industrialized nations, are among the best legislative instruments for attaining high rates of beverage container collection and recycling. A small deposit paid at the time of purchase is repaid when the container is returned, which is one of the main reasons why DRS systems in nations like Norway and Germany achieve return rates above 90%.

Research demonstrates that DRS makes a substantial contribution to the advancement of CE objectives. DRS is transformational, according to Zorpas (2024), since it involves customers directly in the recycling process, lowering waste and guaranteeing that high-quality recyclables are recycled back into production cycles. Similarly, Dempster et al. (2021) discovered that financial motive, system trust, and convenience all affect customer engagement, highlighting the significance of accessibility and design in successful schemes.

The UK offers a helpful case study of the difficulties in implementing DRS. Although the UK has committed to implementing a DRS, discussions over its operational design, cost-effectiveness, and compatibility with current municipal recycling systems are still ongoing (Kükenthal et al., 2023). This illustrates how crucial it is to modify policies to accommodate national waste management systems.

Reshetnikova et al. (2025) point out that Slovakia's creative DRS strategy increased packaging recovery rates and offered a possible template for other emerging economies, such as Ukraine. Their research highlights how integrating DRS with more comprehensive CE tactics might guarantee both financial and environmental advantages.

Effective DRS, which combine robust policy frameworks with consumer interaction, are generally easy to implement, convenient, and widely available, according to best practices from industrialized economies. Adopting DRS for plastics and beverage packaging could benefit Zambia and other African nations by lowering urban litter, boosting recycling sectors, and immediately advancing SDG 12 (responsible consumption and production).

3.1.2.5 Zero-Waste Cities and Municipal Programs

Zero-waste cities are a revolutionary approach to urban sustainability that seeks to increase resource recovery, reduce landfill usage, and integrate the concepts of the CE into local government. Prominent instances include Stockholm, Sweden, where WtE plants supply district heating for thousands of households, and San Francisco, USA, which diverts approximately 80% of its waste from landfills through required composting and recycling.

Zero-waste cities' conceptual underpinnings place a strong emphasis on the systematic integration of recovery, recycling, reuse, and waste prevention. This model, according to Zaman and Lehmann (2011), was developed in response to the fast urbanization of the world and calls for efficient waste management systems that promote long-term economic and environmental resilience. Möslinger et al. (2023) expand on this idea by emphasizing how zero-waste frameworks enable climate-neutral urban futures by tying carbon reduction tactics and CE practices together.

When it comes to growing zero-waste city initiatives, China provides noteworthy insights. According to Meng et al. (2026), pilot initiatives in places like Shenzhen and Shanghai used digital waste tracking systems, stringent regulations, and public involvement to integrate resource efficiency into daily life. Kurniawan et al. (2021) demonstrate how Sukunan in Indonesia effectively implemented a community-driven zero-waste model in Southeast Asia, lowering reliance on landfills through the integration of recycling cooperatives, household-level composting, and circular awareness programs.

Custódio et al. (2024) contend that municipalities play a crucial role in accomplishing zero-waste transitions at the governance level. Their research emphasizes how crucial public–private partnerships and municipal regulations are to coordinate regional waste management plans with more general CE goals. This emphasizes that zero-waste is a governance and policy goal in addition to being technical difficulty.

When combined, best practices from both developed and emerging economies show that communities, robust regulatory frameworks, and local leadership all work together to create successful zero-waste cities. Incorporating zero-waste practices into MSW, such as resource recovery parks, community composting, and required segregation, could greatly lessen urban waste issues in African countries while also advancing sustainable development and climate goals.

3.1.2.6 Public–Private Partnerships

Through the sharing of resources, knowledge, and risks between the public and private sectors, PPPs have become an essential tool for quickening the transition to a CE. PPPs have been crucial in helping developed economies scale circular infrastructure, such as recycling facilities and renewable energy systems, and in encouraging creativity in sustainable urban planning.

PPPs are successful when they include explicit CE models and success determinants. Effective PPP frameworks include circularity ideas into infrastructure design and operations, guaranteeing long-term resource efficiency (Akomea-Frimpong et al. 2024). Adding to this, Akomea-Frimpong et al. (2025) state that to advance CE results in PPP projects, important enablers including stakeholder trust, robust regulatory frameworks, financial incentives, and transparent governance are necessary.

Best practices from throughout the world demonstrate how PPPs can have a transformative effect. According to Vassileva and Simić (2023), green PPPs in Europe have sped up investments in eco-industrial parks, sustainable building, and renewable energy, offering models that other regions can follow. According to Dsilva et al. (2025), despite institutional obstacles, PPPs are being used in the Middle East to prototype CE business models, investigate creative financing options, and address waste management issues.

According to Sharma's (2007) research on Asian e-government partnerships, PPPs have been effectively used for purposes other than waste management. PPPs can provide efficiency and transparency in service delivery. These findings are also applicable to CE governance, where responsibility and cooperation are essential for systemic change. All things considered, PPPs in developed economies show that shared accountability, strong policy frameworks, and cooperative governance are essential for growing CE projects. Building PPPs in sectors like sustainable agriculture, renewable energy, and recycling infrastructure might help developing achieve its green growth objective by accelerating investment, lowering public financial burdens, and creating inclusive green jobs.

3.1.2.7 Policy recommendations for promoting circular economy

Deliberate policy actions that incorporate lessons from industrialized economies and modify them for local contexts are necessary to promote a strong CE. To hold manufacturers responsible for the whole life cycle of their products, governments need first fortify EPR regimes. Policies must encourage innovation in reuse, repair, and recycling by requiring eco-design and premarket accountability in addition to end-of-life management.

Second, by establishing eco-industrial parks where waste from one business is converted into input for another, governments can promote industrial symbiosis. Companies can join resource-sharing networks with incentives like tax exemptions, grants, or preferential land access, and digital platforms can increase the effectiveness of material exchange.

Third, policies need to encourage innovation and circular product design. Product rules should incorporate durability, repairability, and recyclability standards in addition to financing for research to expedite reverse logistics and sustainable product development. Systemic circularity will be ensured by coordinating supply chain management with product policies.

Fourth, to increase recycling rates above 90%, governments should implement DRS for plastics and beverage containers, utilizing consumer incentives. Successful

European models show that DRS not only reduce litter but also ensures that industry receives high-quality recyclables.

Fifth, by requiring resource recovery parks, composting, and waste segregation, governments should be given the authority to lead zero-waste city initiatives. Scaling urban circularity requires strong local leadership backed by national finance and citizen involvement.

Finally, to raise funds and skills for circular infrastructure, PPPs need to be given top priority. Green PPPs can facilitate investments in sustainable agriculture, recycling, and renewable energy while maintaining accountability through open governance and public involvement.

When combined, these measures have the potential to foster an atmosphere in which circularity is incorporated into economic planning, promoting sustainable growth, lowering environmental stressors, and aiding in the accomplishment of both national and international development objectives.

3.2 FINANCIAL AND ECONOMIC INSTRUMENTS

Because they affect how resources are distributed, how markets operate, and how stakeholders act, financial and economic tools are essential for facilitating the adoption of the CE. The adoption of sustainable production and consumption patterns has been hampered in many emerging economies by a lack of fiscal incentives, significant investment risks, and restricted access to financing. Therefore, by promoting investments in waste recovery, recycling, renewable energy, and green innovation, financial and economic instruments work as policy tools that match economic activity with environmental and social goals.

Green bonds, concessional loans, tax breaks, subsidies, eco-levies, carbon pricing, and EPR fees are just a few examples of the many methods that fall under this category. When well designed, they encourage private sector involvement, lessen the financial burden on companies implementing circular methods, and encourage ethical consumer behavior. Additionally, by guaranteeing that polluters pay for the waste they produce and that environmental benefits are monetarily rewarded, they assist governments in internalizing the environmental costs of linear economic operations.

Financial and economic tools are becoming more widely acknowledged in the African setting as drivers of attaining sustainability and national development objectives. Nations including South Africa, Kenya, and Zambia are experimenting with a variety of methods, from public–private partnerships and green credit lines to recycling funding and EPR schemes. These initiatives show an increasing recognition that in addition to regulations, well-designed financial incentives and investment structures are necessary for sustainable transitions to make circular solutions profitable.

Therefore, the creation and expansion of financial and economic tools are essential to furthering the goal of the CE. By bridging the gap between the aspirations of policy and its actual application, they guarantee that financial institutions actively promote equitable economic growth, job creation, and environmental sustainability.

3.2.1 Investments Opportunities and Funding Mechanisms

Financial and economic mechanisms are essential in enabling the transition to CE by mobilizing resources, mitigating investment risks, and establishing conducive market circumstances for sustainable production and consumption. In developing economies, where capital limitations and infrastructural deficiencies frequently obstruct sustainable transitions, these instruments offer essential leverage for enhancing innovation in waste management, renewable energy, water recycling, and sustainable agriculture. Ezeudu and Bristow (2025) assert that strategic finance mechanisms improve resource efficiency and foster financial sustainability by facilitating value recovery and reinvestment in waste valorization. Awino and Apitz (2024) assert that redirecting money flows toward circular models, informed by the waste hierarchy and CE frameworks, is crucial for attaining environmental and socioeconomic objectives.

The Green Growth Strategy and the Eighth National Development Plan (8NDP) of Zambia specifically advocate for circular investments in recycling infrastructure, IS, and renewable energy initiatives as components of the nation's comprehensive sustainable industrialization program. These policy commitments indicate an increasing acknowledgment that green finance can act as a catalyst for inclusive and low-carbon economic development. Similar programs are present throughout Africa: Kenya's Green Economy Strategy and Implementation Plan (GESIP) and South Africa's Green Fund have been pivotal in offering specialized finance for eco-innovation, green entrepreneurship, and circular business models. These instances demonstrate how structured investment frameworks, when combined with legislative incentives, can enhance private sector involvement in the CE transition.

Moreover, research from Nigeria indicates that efficient finance structures are crucial for surmounting systemic obstacles in SWM. Ezeudu and Ugochukwu (2024) emphasize that inadequately coordinated finance mechanisms, restricted budgetary autonomy of municipalities, and insufficient private sector involvement hinder advancements in circular waste recovery. Their findings corroborate previous research by Ezeudu et al. (2021), which identified enablers and barriers to CE implementation in urban markets, highlighting that financial inclusion, accessible credit, and incentives for waste entrepreneurs substantially improve material recovery rates and the integration of the informal sector. These observations correspond with Moh (2017), who noted that Malaysia's advancements in waste recycling and source separation were primarily propelled by financial incentives, subsidy initiatives, and specialized recycling funds that enhanced public involvement and company participation.

PPPs and blended finance have emerged as vital mechanisms for mobilizing money and mitigating the perceived risks linked to circular investments. Regional development banks, including the African Development Bank (AfDB) and the Development Bank of Southern Africa (DBSA), have established specialized green funding windows to facilitate sustainable urban waste management and circular manufacturing projects. Institutional support is augmented by global climate finance instruments, notably the Green Climate Fund (GCF) and the Global Environment Facility (GEF), which offer concessional loans and grants to developing nations executing CE-aligned initiatives. These grants are crucial in closing financial disparities and stimulating domestic resource mobilization.

At the micro level, business initiatives illustrate how private investments can enhance waste value chains while providing social co-benefits. The Manja Pamodzi initiative in Zambia, established by Zambian Breweries, has established an effective system for the collection, aggregation, and recycling of plastic waste. This technique enhances material recovery rates and creates work possibilities for informal waste pickers, thus integrating economic empowerment with environmental stewardship. Comparable inclusive frameworks have been identified in Poland and China, where financial incentives and digitalization-driven CE strategies have expedited transformations in municipal waste systems (Smol et al., 2022; Kurniawan et al., 2022).

The transition of waste and resource systems toward circularity is significantly reliant on innovative financial tools that align environmental concerns with economic incentives. Ezeudu and Bristow (2025) assert that the viability of CE interventions depends on the establishment of financial ecosystems that include government regulation, private sector innovation, and public engagement. By carefully utilizing PPPs, green bonds, recycling funds, and climate finance, developing nations such as Zambia, Kenya, and South Africa can access new investment opportunities that promote industrial diversification, job development, and environmental sustainability.

3.2.2 Economic Incentives for Consumers and Businesses

Economic incentives are essential tools for expediting the shift to a CE, as they encourage behavioral modification and offer financial motivation for both producers and consumers to embrace sustainable methods. In numerous emerging economies, conventional command-and-control strategies have demonstrated inadequacy in fostering long-term transformation due to enforcement difficulties, constrained institutional capacity, and insufficient consumer awareness. As a result, governments are progressively using fiscal and market-oriented instruments such as tax incentives, environmental levies, and EPR frameworks to internalize environmental costs and promote circular market practices (Ezeudu & Bristow, 2025; Awino & Apitz, 2024). These instruments generate a pricing signal that incentivizes circular innovation and penalizes unsustainable practices, harmonizing economic growth with environmental conservation.

The EPR Statutory Instrument No. 65 of 2018 in Zambia requires producers to fund the collection and recycling of postconsumer packaging and other regulated items. This method, aligned with CE concepts, efficiently reallocates waste management expenses from municipalities to manufacturers and consumers who derive advantages from the product life cycle. The Sustainable Waste Management Act of Kenya and the 2024 EPR Regulations include analogous principles, mandating producers to fund take-back systems and collection initiatives for plastics, electronics, and packaging materials. Similarly, South Africa's Section 18 EPR Regulations have formalized the "polluter pays" idea, imposing financial responsibilities on producers to achieve national recovery and recycling objectives. These solutions alleviate the financial strain on local authorities while also fostering eco-design and life cycle accountability throughout value chains (Smol et al., 2020; Kurniawan et al., 2022).

Empirical research substantiates the efficacy of these incentives in enhancing waste recovery and material circularity. Ezeudu et al. (2021) examined urban marketplaces in Nigeria and discovered that informal waste participants have a favorable response to financial incentives, including buyback schemes and material rebates, hence enhancing engagement in waste valorization. Likewise, Moh (2017) emphasized that Malaysia's recycling successes were partially due to stable incentive frameworks, such as DRS, varied garbage collection prices, and subsidies for recyclers, which fostered public engagement in source separation. In advanced environments such as Poland and China, Smol et al. (2020) and Kurniawan et al. (2022) show that fiscal tools, when integrated with digital monitoring systems, enhanced accountability and facilitated data-driven resource recovery insights that can guide policy adaptation in African economies.

Consumers play a vital part in maintaining circular networks, and effectively structured incentives can markedly affect purchasing decisions. DRF for plastic bottles, pay-as-you-throw initiatives, and incentives for reusable packaging have effectively influenced consumer behavior toward responsible purchasing and waste minimization. In Zambia, consumer-oriented efforts within the Manja Pamodzi program have cultivated a recycling culture by connecting community collection stations to recycling firms via financial incentives. This model reflects the conclusions of Ezeudu and Ugochukwu (2024), who contend that accessible incentives and dependable payment methods bolster confidence and engagement among local communities and informal waste workers.

Economic measures aimed at micro, small, and medium-sized enterprises (MSMEs) in the waste and recycling sectors are of equal significance. Microfinance programs, environmentally sustainable credit lines, and concessional loans from development banks like the AfDB and DBSA enable small entrepreneurs to invest in collection, sorting, and processing facilities. Ezeudu and Bristow (2025) assert that inclusive finance methods mitigate inequality and promote entrepreneurship among low-income individuals frequently marginalized by formal credit institutions. When integrated with legislative support and digital traceability technologies, these instruments foster a more transparent and equitable circular market ecology.

Economic incentives serve as drivers for systemic transformation by aligning financial resources with sustainability goals, fostering innovation, and enhancing inclusive economic participation. The efficacy of these instruments relies on coherent regulatory frameworks, consistent enforcement, and collaborative engagement among government, industry, and civil society stakeholders. According to Awino and Apitz (2024) and Ezeudu et al. (2021), sustainable advancement in developing economies hinges on enhancing the integration of fiscal instruments with social inclusion strategies, ensuring equitable distribution of the benefits of circular transitions among producers, consumers, and informal sector participants.

3.3 TECHNOLOGICAL INNOVATIONS

Technological innovation is fundamental to the shift toward a CE and significantly enhances the efficiency, transparency, and sustainability of SWM systems. Through the improvement of waste collection, segregation, processing, and recovery, technology facilitates the transition from conventional linear models of "take–make–dispose" to circular systems wherein resources are reused, recycled, and reintegrated into the production cycle. In developing economies, these innovations are crucial as they provide potential to bridge infrastructure deficits, enhance service delivery, and create inclusive green jobs.

3.3.1 Emerging Technologies in SWM

Emerging technologies are revolutionizing SWM from a conventional linear, disposal-focused system into a dynamic, resource-efficient approach consistent with the concepts of the CE. Innovations in Industry 4.0 such as artificial intelligence (AI), robotics, and the Internet of Things (IoT) waste progressively being incorporated into waste management systems to automate, monitor, and enhance the entire waste value chain. These technologies are transforming the methods of waste collection, sorting, processing, and repurposing, facilitating a transition to data-driven decision-making and improved material recovery.

AI-driven systems can now accurately identify, categorize, and segregate diverse waste fractions, thereby minimizing contamination and enhancing the efficiency of recycling processes. Robotics and sensor-based automation facilitate expedited and safer sorting, reducing human exposure to hazardous garbage. Likewise, IoT-integrated smart bins and collection vehicles utilize real-time sensors to monitor fill levels, track collection routes, and enhance logistics, thereby minimizing fuel consumption and operational expenses. Recent trends in innovation within Industry 4.0-based SWM indicate that these technological solutions enhance waste operations and underpin a digital CE focused on transparency, traceability, and ongoing resource circulation (Cheah et al., 2022; Kanojia & Visvanathan, 2021).

WtE technologies provide an essential element of this shift. Processes including anaerobic digestion, pyrolysis, and gasification transform organic waste into valuable by-products, including biogas, charcoal, and energy. These technologies facilitate the shift to a bioeconomy by reclaiming energy and nutrients from waste streams that would otherwise be disposed of in landfills. Farooq et al. (2021) contend that the incorporation of WtE systems into MSWM policies improves environmental sustainability and energy security, especially for cities experiencing landfill saturation and significant reliance on fossil fuels. In developing economies, small- and medium-sized biogas facilities and biomass conversion systems are demonstrating efficacy in mitigating local energy deficits while decreasing greenhouse gas emissions.

Technological innovation in waste management is rapidly advancing across Africa as cities strive to upgrade existing systems and enhance service delivery. Kenya and South Africa have developed advanced recycling facilities that utilize automated sorting and energy recovery technology to optimize material use and reduce landfill waste. Zambia's Eighth National Development Plan (8NDP) has prioritized investment in waste valorization, eco-industrial parks, and circular manufacturing hubs. These facilities utilize a combination of mechanical, biochemical, and thermal processes to convert plastics, metals, and organic waste into commercially viable products, including compost, fuel briquettes, and recycled plastic granules.

The implementation of these technologies not only yields environmental advantages but also fosters entrepreneurship, enhances skill development, and generates employment opportunities. Mandpe et al. (2023) emphasize that the integration of CE strategies with technology innovation enables developing economies to enhance local value chains by promoting inclusive business models that involve informal waste pickers, small recyclers, and community-based firms. This integration of technology and circular thinking provides a means for emerging nations to surpass traditional waste management methods, fostering adaptive and resilient urban economies propelled by innovation and sustainability.

Emerging technologies in solid waste management highlight the practical implementation of Industry 4.0 and CE synergies. Through the integration of automation, digitization, and energy recovery advances, nations may convert garbage from a liability into a strategic asset, promoting environmental conservation, economic diversification, and social inclusion concurrently.

3.3.2 Digital Solutions for SWM and Tracking

Digitalization has become one of the most transformative forces in advancing efficient, transparent, and circular SWM systems. As the world embraces Industry 4.0, digital technologies are redefining how waste is generated, collected, tracked, and valorized. Tools such as mobile applications, blockchain systems, the IoT, and Geographic Information Systems (GIS) are enabling data-driven decision-making and real-time visibility across the waste management value chain. These innovations enhance efficiency, reduce operational costs, and improve accountability among waste generators, collectors, recyclers, and regulators (Cheah et al., 2022; Kurniawan et al., 2022).

Modern waste management increasingly relies on smart technologies that integrate sensors and communication networks to optimize resource use and logistics. IoT-enabled smart bins, for example, can monitor fill levels and automatically signal collection vehicles when servicing is required, preventing overflows and unnecessary trips. Similarly, GPS and GIS tools allow city councils and private operators to map waste generation hotspots, track fleet performance, and identify inefficiencies in collection routes. This digital infrastructure facilitates predictive maintenance, reduces fuel consumption, and supports evidence-based policymaking. Sosunova and Porras (2022) describe such systems as foundational to "smart cities," where waste management is part of an interconnected urban ecosystem powered by AI, cloud computing, and real-time analytics.

Blockchain technology is also emerging as a powerful tool for enhancing transparency and trust within waste value chains. By recording each transaction and transfer of materials on immutable digital ledgers, blockchain ensures the traceability of waste from its source to its destination. This promotes producer accountability, reduces illegal dumping, and supports compliance with environmental regulations. Castiglione et al. (2023) proposes a blockchain-based framework that integrates producers, waste collectors, and recyclers within a digital CEnetwork, ensuring that materials and financial flows are tracked transparently and that incentives are distributed fairly among stakeholders. Such systems can strengthen extended producer responsibility (EPR) implementation by providing verifiable data on material recovery, recycling rates, and compliance performance.

Across developing regions, particularly in Africa and Latin America, digital innovations are enabling leapfrogging opportunities for circularity. Rihm et al. (2024) highlight how digital transformation is driving innovation in waste collection and recycling in Latin American cities through mobile platforms, business intelligence systems, and data sharing tools that connect households, collectors, and recyclers. These systems often feature reward mechanisms or e-payment options that incentivize citizens to separate and return recyclables. Similar trends are emerging in Africa, where start-ups and municipalities are adopting mobile-based platforms that link waste generators to recycling enterprises through real-time communication and payment integration.

In Zambia, the government's Smart Zambia Master Plan underscores the importance of digital transformation as part of its broader green growth and CE agenda. IoT-enabled tracking systems are being explored to monitor waste volumes, collection efficiency, and recycling rates across municipalities. These digital initiatives enhance coordination between local authorities and private recyclers, while generating reliable data for planning and performance evaluation. The integration of such systems provides a foundation for evidence-based policymaking, efficient allocation of resources, and compliance monitoring—addressing one of the persistent gaps in traditional SWM systems: the lack of accurate, real-time data.

Digitalization also facilitates the inclusion of informal waste actors in formal waste economies. By providing digital identification, e-payment options, and traceable transaction records, small recyclers and waste pickers can gain formal recognition, fair compensation, and access to new markets. This inclusion not only improves livelihoods but also strengthens the social dimension of CE transitions. Digital systems that capture the contributions of informal workers make them visible participants in sustainable waste ecosystems—transforming a historically marginalized group into active agents of circularity.

The potential of digital transformation in SWM extends beyond operational efficiency to system-wide innovation. Business intelligence platforms can aggregate data from multiple stakeholders, providing insights into material flows, consumer behavior, and recycling performance. These data-driven insights inform strategic interventions such as investment planning, regulatory reform, and infrastructure development. When combined with IoT, AI, and blockchain technologies, digitalization creates an integrated circular waste ecosystem that is self-monitoring, adaptive, and transparent.

In essence, digital technologies are not only optimizing the mechanics of waste management but also redefining governance, accountability, and stakeholder participation. As demonstrated across Asia, Latin America, and Africa, digital tools make it possible to transform waste systems into intelligent networks that close resource loops, empower communities, and support sustainable urban development. For developing economies like Zambia, embracing these innovations is essential for achieving both environmental and socioeconomic resilience in the CE era.

3.3.3 Technology Adaptation for Developing Economies

In numerous emerging economies, the problem of attaining sustainable SWM extends beyond mere access to advanced technologies; it involves the adaptation of these technologies to local economic, environmental, and social contexts. Technologies developed for industrialized nations are frequently excessively costly, difficult to maintain, and incompatible with the infrastructural and institutional capabilities of lower-income countries. Consequently, the emphasis should transition from simple technology adoption to contextually tailored technological adaptation, guaranteeing that innovation is cost-effective, scalable, and pertinent to local conditions (Khan & Ali, 2022).

Developing nations are poised to gain the most from affordable, decentralized, and community-oriented waste management technology that may be incorporated into current informal systems. Examples encompass small-scale composting facilities, solar-powered garbage compactors, low-emission incinerators, and locally manufactured recycling apparatuses such as plastic shredders, balers, and briquette presses. These technologies are simpler to construct and sustain, frequently depend on locally sourced resources, and can be managed by small businesses or community cooperatives. In addition to their technological benefits, such improvements promote local production, foster entrepreneurship, and diminish reliance on imported machinery.

An essential element of effective technology adaptation is capacity building the ongoing enhancement of local skills, knowledge, and institutional structures that facilitate technology design, maintenance, and innovation. Collaborations among governments, academic institutions, and commercial firms are pivotal in this process. Universities in Zambia, Kenya, and South Africa are partnering with municipalities and small firms to develop and prototype waste processing technology tailored for informal sector application. These collaborations facilitate the connection between research and practical application, cultivating innovation ecosystems where students, engineers, and waste entrepreneurs together develop solutions for localized CE models.

Training initiatives in domains such as digital literacy, mechanical repair, and equipment maintenance enable local populations to assume control of modern technology. This guarantees both operational sustainability and social inclusion, enabling informal waste pickers frequently marginalized from advanced systems to actively engage in enhanced waste value chains. Yadav, Soni, and Kumar (2023) assert that inclusive strategies are vital for surmounting obstacles that impede the implementation of smart waste management systems in developing regions, such as inadequate finance, policy inconsistency, and a lack of skilled personnel.

Policy support and tailored financial channels are both essential for technology adaptation. Governments can expedite adoption by offering tax incentives, innovation grants, and green finance mechanisms that render sustainable solutions accessible to small-scale recyclers and local manufacturers. Incorporating these instruments into national CE initiatives guarantees that technological implementation supports overarching development objectives, including industrial diversification, poverty alleviation, and job creation. This convergence signifies the increasing acknowledgment that advancements in waste management must yield triple bottom line advantages in environmental preservation, economic enhancement, and social integration.

Moreover, the adaption of technology should adhere to the waste hierarchy principles, emphasizing reduction, reuse, and recycling rather than disposal. Awino and Apitz (2024) assert that sustainable waste management frameworks in poor nations must prioritize innovations that diminish material intensity and improve the recovery of secondary resources. Cost-effective material recovery facilities, biodegradable waste digesters, and mobile recycling units exemplify how developing economies may incorporate circular economy-oriented technology without straining existing infrastructures.

The experiences of African cities underscore the significance of incremental and adaptive innovation. Kenya's nascent recycling enterprises, South Africa's community-based waste-to-energy efforts, and Zambia's eco-industrial ventures exemplify the efficacy of locally suitable technology in realizing significant environmental and socioeconomic benefits. By leveraging indigenous knowledge and local entrepreneurship, these systems foster technical sovereignty, enabling developing countries to create for their specific settings instead of solely relying on imported solutions.

Ultimately, successful technology adaption necessitates a comprehensive framework that integrates technical innovation, institutional collaboration, and community engagement. This methodology guarantees that emerging technologies are both ecologically sustainable and economically viable, as well as socially inclusive. With cohesive regulatory frameworks and ongoing capacity-building activities, emerging economies may convert waste management from a reactive service into a proactive catalyst for circular, resilient, and equitable growth.

3.4 COMMUNITY ENGAGEMENT AND AWARENESS

The shift to a CE in SWM essentially relies on community awareness, active engagement, and ownership of waste-related practices. Evidence from Zambia and global research indicates that communities must serve as active participants, rather than passive beneficiaries, for community engagement efforts to thrive. The subsequent subsections delineate comprehensive engagement tactics substantiated by empirical research.

3.4.1 Strategies for Community Engagement and Awareness

3.4.1.1 Participatory Waste Governance

Participatory governance guarantees community involvement in the design, decision-making, and oversight of waste management systems. Izdebska and Knieling (2020) assert that citizen engagement bolsters legitimacy, improves compliance, and matches waste management methods with local conditions. In Chelstone Township, Chisanga et al. (2024) discovered that households had a greater propensity to sort waste and embrace sustainable practices when engaged through neighborhood committees, community forums, and stakeholder dialogues.

Global experiences further substantiate this assertion. Chachami-Chalioti et al. (2024) indicate that in Northern Greece, CE activities integrated into the national waste management plan proved more effective when residents engaged in establishing local recycling objectives and assessing system efficacy. This collaborative method fosters trust, enhances local ownership, and ensures the sustainability of CE adoption post-project interventions.

3.4.1.2 Tailored Awareness and Behaviour Change Campaigns

Awareness deficiencies are a significant obstacle to the adoption of CE practices. Mwanza and Mbohwa (2019) emphasize that in Zambia, public awareness regarding the effects of plastic waste, recycling possibilities, and suitable disposal techniques is insufficient. Customized awareness campaigns by utilizing radio, churches, market sensitization, ward meetings, and social media facilitate the contextualization of CE practices for local populations.

Chareonvong et al. (2025) affirm that culturally appropriate behavior change communication, conveyed in local languages, enhances the adoption of waste segregation and recycling practices. In the Thai context, Sridan and Surapolchai (2020) demonstrate that continual awareness campaigns, bolstered by local leaders, lead households to transition from passive waste disposal to active engagement in CE practices, including composting and reuse. Effective communication enables communities to perceive CE not merely as an abstract notion but as a pragmatic solution that diminishes expenses, enhances health, and generates revenue opportunities.

3.4.1.3 School-Based and Youth-Focused CE Education

Education serves as a pivotal function in cultivating a sustainable culture of continuous improvement. Chareonvong et al. (2025) demonstrate that the incorporation of CE principles into educational curricula, environmental clubs, and youth innovation initiatives cultivates early awareness and promotes enduring sustainable practices. Adolescents frequently impact domestic behaviors by advocating for waste segregation, deterring littering, and endorsing upcycling initiatives.

In Zambia, civic education inside schools and youth organizations, including church youth ministries and community clubs, provides an economical means for widespread integration of civic education knowledge. Global evidence further corroborates this methodology. CE educational projects in Greece and Thailand illustrate that youth-led campaigns markedly enhance household engagement in recycling and composting (Sridan & Surapolchai, 2020; Chachami-Chalioti et al., 2024). Engaging the youth thereby expedites generational transitions toward circular lives.

3.4.1.4 Community Recycling, Demonstration, and Innovation Hubs

Functional learning environments allow communities to observe the implementation of CE principles. These hubs may encompass community recycling centers, repair cafés, composting demonstration sites, and upcycling workshops. Chareonvong et al. (2025) assert that these facilities function as "living laboratories" in which people acquire sorting procedures, concepts of reuse, and the economic significance of recyclable items.

Case studies from Asia and Europe illustrate that hubs enable households to comprehend the advantages of the CE, transforming theory into practical skills. In Northern Greece, the establishment of community drop-off locations and repair centers enhanced local recycling rates and promoted the reuse of electronics and domestic goods (Chachami-Chalioti et al., 2024). In Zambia, informal recycling sites exist but necessitate governmental acknowledgment, enhanced infrastructure, and community stewardship to transform into effective CE innovation hubs.

3.4.1.5 Integration of Informal Waste Pickers and Local Champions

Informal waste pickers are essential participants in Africa's waste recovery system. Mwanza and Mbohwa (2019) and Chisanga et al. (2024) note that they play a crucial role in material recovery, although frequently function without acknowledgment or assistance. Incorporating waste pickers as ambassadors for the CE bolsters community trust, amplifies waste diversion, and sustains livelihoods.

Global evidence highlights comparable results: Izdebska and Knieling (2020) contend that CE solutions are more efficacious when they explicitly incorporate informal participants with substantial recycling expertise. Sridan and Surapolchai (2020) illustrate that coordination between local authorities and informal recyclers in Thailand enhanced system efficiency and household compliance. When informal participants engage as instructors and collection collaborators, communities acquire a more comprehensive awareness of recycling processes and the economic advantages of circularity.

3.4.1.6 Multi-stakeholder Collaboration and Local Leadership

The adoption of CE principles flourishes when local leaders vigorously advocate for sustainable practices. Indigenous leaders, council members, clergy, educators, and local entrepreneurs may galvanize communities more efficiently than external agents. Chisanga et al. (2024) demonstrate that local leadership endorsement markedly enhanced participation in waste management efforts in Chelstone Township.

Chachami-Chalioti et al. (2024) similarly discovered that the adoption of CE in Northern Greece enhanced when municipalities, NGOs, and citizen groups together established waste management initiatives. Collaboration across multiple stakeholders guarantees resource alignment, uniform messaging, and ongoing momentum. Local leaders facilitate the institutionalization of CE activities through bylaws, cleanup schedules, and public acknowledgment of households that engage in recycling.

3.4.1.7 Systemic and Community-Centered Integrated Waste Management

Sridan and Surapolchai (2020) underscore a systems-thinking perspective, positioning community participation as an integral component of a comprehensive waste management system. This approach integrates households, local governments, informal recyclers, educational institutions, and corporations into a unified CE framework. Communities are educated not only on the measures to undertake but also on the significance of those acts within the broader environmental and economic framework.

This methodology facilitates the transition from linear "collect–transport–dispose" behaviors to circular activities, including reuse, repair, composting, and recycling. It enables communities to observe the immediate effects of their waste practices on local cleanliness, public health, and revenue generation, thus strengthening the long-term adoption of CE principles.

3.4.2 Strategies for Influencing Public Participation

Public engagement is widely acknowledged as a fundamental element of effective CE execution in SWM systems. In various contexts, including Zambia, Europe, and Southeast Asia, evidence indicates that the engagement of residents, households, youth, informal actors, and local organizations fosters social ownership, enhances compliance, and expedites the shift from linear waste behaviors to circular practices.

3.4.2.1 Enhancing Knowledge, Awareness, and Behavioral Change

Public participation enables communities to comprehend the environmental, economic, and health consequences of unsustainable waste practices. Mwanza and Mbohwa (2019) identified that insufficient knowledge and awareness regarding plastic waste management in Zambia are significant obstacles to the development of a CE. Active public engagement via meetings, campaigns, and educational programs enhances people's comprehension of the significance of waste sorting, reuse, recycling, and material recovery.

Chareonvong et al. (2025) assert that community-engaged CE education fosters tangible behavioral modifications, such as regular household waste segregation and conscientious consumption practices. Participation thus transforms waste management from a government-driven endeavor to a collective societal obligation.

3.4.2.2 Strengthening Local Ownership and Acceptance of CE Initiatives

CE solutions are more efficient and sustainable when communities participate in the design, planning, and oversight of waste systems. Public participation fosters local ownership, guaranteeing that CE actions correspond with cultural norms, home circumstances, and community priorities.

Chisanga et al. (2024) illustrate that in Chelstone Township, households exhibit greater willingness to engage in waste segregation and collection programs when given the opportunity to submit ideas, express concerns, and participate in decision-making processes. This sense of ownership diminishes opposition, enhances compliance, and prolongs the sustainability of waste management projects.

Izdebska and Knieling (2020) demonstrate that European towns employing participatory methods attain elevated recycling rates, as citizens perceive themselves as co-creators in CE systems rather than passive recipients.

3.4.2.3 Improving Policy Effectiveness and Service Delivery

Public engagement enhances the congruence between policy objectives and community requirements. Through community engagement, politicians and local authorities acquire understanding of tangible obstacles, such as insufficient recycling bins, inconsistent garbage collection, and restricted market access for recyclables.

Chachami-Chalioti et al. (2024) demonstrate that public feedback in Northern Greece enabled municipalities to optimize garbage collection routes, enhance communication campaigns, and fortify CE components in the national waste management plan. In Zambia, this engagement guarantees that CE plans align with local conditions, including urban informal settlements, dependence on informal garbage collectors, and inadequate infrastructure.

In the absence of community engagement, CE policies frequently suffer from inadequate implementation or neglect, while being well conceived.

3.4.2.4 Building Trust and Strengthening Multi-stakeholder Collaboration

Engagement cultivates trust between communities and authorities, facilitating more efficient collaboration in waste management. Sridan and Surapolchai (2020) emphasize that community engagement in Thailand fostered robust partnerships among people, local leaders, informal recyclers, and municipal authorities. This relationship facilitated more efficient coordination of collection schedules, neighborhood clean-ups, composting initiatives, and the establishment of recycling groups.

Chisanga et al. (2024) demonstrate that in Zambia, community gatherings, door-to-door sensitization, and the involvement of local leaders cultivate trust and promote household collaboration with waste management initiatives. Participation grounded by trust mitigates skepticism, diminishes misinformation, and fortifies the social framework essential for enduring community engagement initiatives.

3.4.2.5 Increasing Material Recovery Rates and Supporting Circular Value Chains

Public engagement directly enhances both the volume and caliber of recyclable materials entering the circular value chain. Household involvement increases the likelihood of source segregation of garbage, hence minimizing contamination and reducing recycling expenses.

Chareonvong et al. (2025) discovered that active public engagement in CE education resulted in markedly increased material recovery rates. Izdebska and Knieling (2020) assert that citizen participation is crucial for urban circularity, aiding cities in diminishing waste-to-landfill quantities and enhancing recycling process efficiency.

In Zambia, informal waste pickers, reliant on material recovery for their livelihoods, gain from the enhanced quality of waste streams produced via public engagement. This fortifies circular value networks and amplifies economic prospects for marginalized populations.

3.4.2.6 Ensuring Social Inclusion and Equity in Waste Systems

Engagement guarantees that CE activities are inclusive and socially attuned. Numerous communities, particularly in economically disadvantaged urban regions, rely on informal waste management systems. Engagement facilitates the identification of the requirements of at-risk households, women, youth, and informal collectors, guaranteeing that CE policies do not exclude or marginalize these groups.

Research conducted by Chisanga et al. (2024) indicates that engaged communities can recognize social impediments, like restricted access to garbage bags and insufficient household space for waste storage. Izdebska and Knieling (2020) contend that citizen-excluding CE systems generate disparities and inefficiencies, whereas participatory models foster equity and social harmony.

3.4.2.7 Supporting Long-Term Sustainability and Behavioral Institutionalization

Public engagement facilitates the integration of CE principles into everyday practices, cultural standards, and communal customs. When communities engage consistently via awareness campaigns, educational programs, and local leadership, circular behaviors integrate into daily life.

Chareonvong et al. (2025) assert that the enduring success of CE is contingent upon the institutionalization of behaviors over generations. Likewise, Sridan and Surapolchai (2020) note that communities exhibiting robust participatory cultures preserve sustainable waste management practices post the conclusion of foreign initiatives. This emphasizes public engagement as a sustained investment in sustainability.

3.4.3 Strategies for Changing Behaviors

Behavioral change and public awareness are essential catalysts for the implementation of CE, especially in community SWM. Data from both developing and developed environments indicates that individuals' knowledge, attitudes, and behaviors substantially affect waste prevention, reuse, sorting, recycling, and responsible consumption. The subsequent strategies are identified in the literature as effective methods for altering behaviors and enhancing awareness.

3.4.3.1 Community-Centered CE Education and Structured Learning Platforms

Education is crucial in enabling communities to embrace CE principles. Chareonvong et al. (2025) assert that educational programs centered on CE principles, administered via schools, community centers, and local government initiatives, establish a fundamental comprehension of waste reduction, reuse, and recycling methodologies. Their comprehensive evaluation emphasizes that interactive training, demonstrations, and practical exercises enhance understanding and promote the sustained adoption of circular practices.

Parajuly et al. (2020) emphasize that focused educational interventions concerning particular waste streams, such as e-waste in their research, successfully modify disposal behaviors by rectifying misconceptions and enhancing knowledge of environmental consequences. Structured learning platforms facilitate the transition from awareness to action by imparting practical skills, like sorting, repairing, and upcycling.

3.4.3.2 Behavior Change Communication and Social Awareness Campaigns

Behavior modification necessitates more than just information; it demands purposeful communication that aligns with individuals' values, emotions, and social identities. Etim (2024) emphasizes that effectively crafted public awareness programs induce behavioral changes when they incorporate emotional appeals, accessible storylines, and practical directions. Such efforts are crucial to entrepreneurial waste management models, wherein public engagement directly influences resource recovery and corporate sustainability.

Puvača et al. (2025) further illustrate that good communication profoundly impacts citizens' perspective of waste as a resource instead of a liability. Within the EU framework, continuous engagement via media, social platforms, and community activities has fostered a culture of recycling and responsible consumption. Communications highlighting the economic benefits of circularity such as employment generation, revenue from recyclables, and cost reductions typically resonate profoundly with communities.

3.4.3.3 Use of Social Norms, Role Models, and Community Champions

Individuals are more inclined to embrace CE behaviors when they observe their peers in the community engaging in similar practices. Parajuly et al. (2020) emphasize the significance of social norms and peer influence in fostering appropriate waste disposal practices, especially for intricate waste categories such as electronic waste.

Community leaders, youth advocates, educators, religious figures, and informal waste collectors can act as exemplary examples for CE practices. Etim (2024) observes that entrepreneurs in waste management frequently serve as prominent catalysts for change, illustrating the economic and ecological advantages of circularity and inspiring community engagement. Social modeling is particularly effective in households with limited prior knowledge or misunderstandings on waste management.

3.4.3.4 Incentives, Rewards, and Community-Based Motivational Mechanisms

Positive reinforcement can expedite behavioral modification by rewarding preferred acts. Puvača et al. (2025) underscore the significance of incentives—such as recycling credits, discounts, municipal prizes, or recognition programs in encouraging residents to engage in CE-related activities. Illustrations encompass recycling-for-points initiatives, contests for the tidiest neighborhoods, certificates of reward for educational institutions, and income-dependent incentives associated with segregated garbage collection.

Etim (2024) observes that financial incentives associated with entrepreneurial waste initiatives promote public engagement, particularly in low-income environments where households pursue supplementary income options.

3.4.3.5 Enhancing Access to CE Infrastructure and User-Friendly Systems

Behavioral change is enhanced when communities possess an accessible and well-structured CE infrastructure. Parajuly et al. (2020) assert that deficiencies in infrastructure frequently limit behavior, despite elevated levels of knowledge. Communities are more inclined to engage in CE initiatives when they have access to recycling bins, collection sites, repair facilities, and electronic waste drop-off locations.

Puvača et al. (2025) demonstrate that enhancements in EU infrastructure such as color-coded bins, accessible recycling centers, and digital waste platforms have markedly increased citizens' recycling behaviors. Efficient infrastructure diminishes behavioral obstacles and renders circular processes instinctive.

3.4.3.6 Embedding CE in School Curricula and Youth Empowerment Programs

Enduring behavioral transformation commences with subsequent generations. Chareonvong et al. (2025) emphasize that educational institutions serve as a strategic foundation for early instillation of CE principles. Incorporating CE principles into disciplines like physics, civics, and environmental studies empowers students to impact domestic behaviors and foster intergenerational transformation.

Programs for youth engagement such as eco-clubs, innovation challenges, and waste art competitions encourage behavioral changes through experiential learning. These programs emphasize that waste constitutes a resource and fosters the development of future circular leaders.

3.4.3.7 Leveraging Digital Tools, Apps, and Online Learning

Digital platforms can augment awareness and shape behaviors via mobile notifications, instructional videos, e-learning modules, and community WhatsApp groups. Parajuly et al. (2020) underscore the growing efficacy of digital interventions in enhancing knowledge of e-waste management, particularly among urban youth, even if not universally emphasized in all research.

Digital campaigns provide economical methods to engage extensive audiences and can tailor messages for various demographics, so amplifying their effectiveness.

3.4.3.8 Multi-stakeholder Collaboration for Coordinated Messaging

Behavioral change is more efficacious when supported by several entities' government, NGOs, corporations, educational institutions, and community organizations. Etim (2024) observes that collaborative business models thrive when all stakeholders communicate similar signals that bolster circular processes.

Puvača et al. (2025) demonstrate that European CE projects achieve optimal effectiveness when underpinned by integrated policy frameworks and harmonized communication tactics. Collaboration among multiple stakeholders guarantees continuity, prevents conflicting signals, and bolsters public trust.

3.5 CASE STUDIES OF SUCCESSFUL COMMUNITY INITIATIVES

3.5.1 Community-Based Circular Waste Management in Tan Deaw Sub-district, Thailand

The Tan Deaw sub-district in Saraburi Province, Thailand, exemplifies how participatory involvement may effectively establish CE programs in SWM. Employing participatory action research (PAR) and soft systems approach, residents, local leaders, and municipal authorities collaboratively developed an integrated waste management strategy based on CE concepts. The intervention focused on residential waste segregation, composting of organic materials, and the creation of a community waste bank for the exchange of recyclables for cash or commodities. These community-driven educational initiatives developed practical competencies, heightened awareness, and enabled residents to perceive garbage as a precious asset. The community established an effective circular system that enhanced material recovery and improved environmental cleanliness, illustrating how co-creation and local ownership may promote sustainable waste practices (Sridan & Surapolchai, 2020).

3.5.2 Household Participation in Chelstone Township, Lusaka, Zambia

Chelstone Township in Lusaka exemplifies the advancements and constraints of community involvement in promoting community engagement-based SWM. The municipality implemented community-based enterprises (CBEs) and decentralized frameworks designed to enhance household involvement in waste management services. Despite sensitization initiatives and public–private collaborations intended to improve source separation and appropriate disposal, participation was inconsistent. Households were infrequently involved in decision-making and voiced apprehensions regarding inconsistent garbage collection, insufficient communication, and a lack of transparency. Nonetheless, the study underscores that even minimal participation enhanced understanding of appropriate waste management and cultivated community interest in more inclusive CE practices. The findings emphasize that profound and direct community engagement beyond simply acquiring information is crucial for attaining sustainable, community-driven CE outcomes (Chisanga et al., 2024).

3.5.3 Reward Recycling and Pay-As-You-Throw in Northern Greece

Northern Greece exemplifies the success of policy-driven CE projects when bolstered by strong citizen engagement. In accordance with the nation's revised national SWM strategy, municipalities have implemented pay-as-you-throw (PAYT) rates, reward recycling (RR) systems, and decentralized Green Points for citizens to deposit sorted waste. A poll of more than 1,600 locals indicated that while initial awareness of these instruments was minimal, community members showed a robust readiness to adopt these practices when offered explicit education and direction. Targeted awareness programs, especially those focused on youth, were demonstrated to be essential in enhancing comprehension of CE tools, boosting confidence in recycling systems, and fostering enduring behavioral change. This case illustrates that CE strategies are most effective when accompanied by ongoing public communication that facilitates active and meaningful citizen participation (Chachami-Chalioti et al., 2024).

REFERENCES

African Development Bank (Af DB) (2022) *Green Growth Framework for Africa*. Abidjan: AfDB.

Akomea-Frimpong, I., Jin, X., Osei Kyei, R., Tetteh, P.A., Tumpa, R.J., Ofori, J.N.A. and Pariafsai, F. (2024) 'A review of circular economy models and success factors in public–private partnership infrastructure development', *Built Environment Project and Asset Management*, 14(1), pp. 109–126. https://doi.org/10.1108/BEPAM-01-2023-0014

Akomea-Frimpong, I., Ofori, J.N.A., Botchway, B., Tumpa, R.J. and Tetteh, P.A. (2025) 'Key success factors towards circular economy in public–private partnership projects', *Smart and Sustainable Built Environment*, pp. 1–27. https://doi.org/10.1108/SASBE-02-2025-0074

Awino, F.B. and Apitz, S.E. (2024) 'Solid waste management in the context of the waste hierarchy and circular economy frameworks: An international critical review', *Integrated Environmental Assessment and Management*, 20(1), pp. 9–35.

Baldassarre, B., Schepers, M., Bocken, N., Cuppen, E., Korevaar, G. and Calabretta, G. (2019) 'Industrial symbiosis: Towards a design process for eco-industrial clusters by integrating circular economy and industrial ecology perspectives', *Journal of Cleaner Production*, 216, pp. 446–460. https://doi.org/10.1016/j.jclepro.2019.01.091

Burke, H., Zhang, A. and Wang, J.X. (2023) 'Integrating product design and supply chain management for a circular economy', *Production Planning & Control*, 34(11), pp. 1097–1113. https://doi.org/10.1080/09537287.2022.2059830

Castellet-Viciano, L., Hernández-Chover, V., Bellver-Domingo, Á. and Hernández-Sancho, F. (2022) 'Industrial symbiosis as a mechanism for implementing circular economy practices', *Sustainability*, 14(23), p. 15872. https://doi.org/10.3390/su142315872

Castiglione, A., Cimmino, L., Di Nardo, M. and Murino, T. (2023) 'A framework for achieving a circular economy using blockchain technology in sustainable waste management systems', *Computers & Industrial Engineering*, 180, p. 109263.

Cecchin, A., Salomone, R., Deutz, P., Raggi, A. and Cutaia, L. (2020) 'Relating industrial symbiosis and circular economy to sustainable development', in Cecchin, A. et al. (eds.) *Industrial Symbiosis for the Circular Economy*. Cham: Springer, pp. 1–25. https://doi.org/10.1007/978-3-030-36660-5_1

Chachami-Chalioti, E., Skourtos, M., Damigos, D. and Goulas, A. (2024) 'Citizen awareness and behavioural responses to pay-as-you-throw and reward recycling schemes', *Waste Management*, 171, pp. 132–144.

Chareonvong, S., Rattanapan, C., Kerdsuwan, S. and Chavalparit, O. (2025) 'Community-based circular economy education for sustainable municipal solid waste management', *Journal of Cleaner Production*, 418, p. 139890.

Cheah, C.G., Chia, W.Y., Lai, S.F., Chew, K.W., Chia, S.R. and Show, P.L. (2022) 'Innovation designs of Industry 4.0-based solid waste management', *Environmental Research*, 213, p. 113619.

Chisanga, B., Mbohwa, C. and Mwanza, B.G. (2024) 'Community-based enterprises and household participation in solid waste management in Lusaka, Zambia', *Sustainable Cities and Society*, 102, p. 105126.

Custódio, M.M., Junqueira, J.C. and Manrique, J.I.T. (2024). 'Zero waste for circular economy: The importance of municipalities and their public policies for circular economy', *Veredas do Direito*, 21, p. e212656.

Dempster, M., Orr, K. and Berry, E. (2021) 'Recycling and deposit return schemes: A consumer survey', *Resources, Conservation and Recycling*, 169, p. 105529.

Den Hollander, M.C., Bakker, C.A. and Hultink, E.J. (2017) 'Product design in a circular economy', *Journal of Industrial Ecology*, 21(3), pp. 517–525.

Domenech, T., Bleischwitz, R., Doranova, A., Panayotopoulos, D. and Roman, L. (2019) 'Mapping industrial symbiosis development in Europe', *Resources, Conservation and Recycling*, 141, pp. 76–98.

Dsilva, J., Usmani, F.Z.A. and Shaikh, M.I. (2025) 'Leveraging public–private partnerships to drive circular economy', in Agarwal, R. et al. (eds.) *Rethinking Resources*. Singapore: Springer, pp. 335–354.

Etim, E.E. (2024) 'Behaviour change communication and entrepreneurship-driven solid waste management', *Journal of Environmental Management*, 350, p. 119725.

Ezeudu, O.B. and Bristow, D. (2025) 'Financing methods for solid waste management', *Sustainable Development*, 33(2), pp. 3062–3085.

Ezeudu, O.B. and Ugochukwu, U.C. (2024). 'Financing mechanism for solid waste management in Anambra, Nigeria: Analyses of emerging challenges and implication for circular economy', *Environmental Science and Pollution Research*, 31(19), pp. 27634–27652.

Ezeudu, O.B., Ezeudu, T.S., Ugochukwu, U.C., Agunwamba, J.C. and Oraelosi, T.C. (2021) 'Enablers and barriers to circular economy implementation', *Environmental and Sustainability Indicators*, 12, p. 100150.

Farooq, A., Miller, K.A., Goh, Y.M. and Aad, G. (2021) 'A framework for the selection of suitable waste-to-energy technologies for municipal solid waste management', *Frontiers in Sustainability*, 2, Article 681690. https://doi.org/10.3389/frsus.2021.681690

Government of Kenya (2014) *National Solid Waste Management Strategy*. Nairobi: Ministry of Environment.

Government of Kenya (2022) *Sustainable Waste Management Act No. 31 of 2022*. Nairobi: Government Printer.

Government of Rwanda (2019) *Law No. 17/2019 on Single-Use Plastics*. Kigali: Government of Rwanda.

Government of South Africa (2008) *National Environmental Management: Waste Act (Act No. 59 of 2008)*. Pretoria: Government Printer.

Government of South Africa (2020) *National Waste Management Strategy 2020*. Pretoria: DFFE.

Government of South Africa (2021) *Extended Producer Responsibility Regulations*. Pretoria: Government Gazette No. 43879.

Government of Tanzania (2019) *Environmental Management (Prohibition of Plastic Carrier Bags) Regulations*. Dar es Salaam.

Government of Zambia (2004) *National Solid Waste Management Strategy*. Lusaka: Environmental Council of Zambia. https://www.unep.org/resources/report/national-solid-waste-management-strategy-zambia?utm_source=chatgpt.com

Government of Zambia (2018a) *Solid Waste Regulation and Management Act No. 20 of 2018*. Lusaka: Government Printer. https://www.parliament.gov.zm/sites/default/files/documents/acts/The%20Solid%20Waste%20Regulation%20and%20Management%20Act%2C%202018.pdf?utm_source=chatgpt.com

Government of Zambia (2018b) *Environmental Management (Extended Producer Responsibility) Regulations, SI No. 65 of 2018*. Lusaka: Government Printer. https://www.parliament.gov.zm/sites/default/files/documents/acts/The%20Solid%20Waste%20Regulation%20and%20Management%20Act%2C%202018.pdf?utm_source=chatgpt.com

Government of Zambia (2022) *Eighth National Development Plan (2022–2026)*. Lusaka: Ministry of Finance.

Izdebska, O. and Knieling, J. (2020) 'Citizen involvement in waste management and circular economy in cities: Key elements for planning and implementation', *European Spatial Research and Policy*, 27(2), pp. 115–129. https://doi.org/10.18778/1231-1952.27.2.08

Kanojia, A. and Visvanathan, C. (2021) 'Industry 4.0 interventions in urban solid waste management', *Waste Management & Research*, 39(11), pp. 1414–1426.

Khan, F. and Ali, Y. (2022). 'A facilitating framework for a developing country to adopt smart waste management in the context of circular economy', *Environmental Science and Pollution Research*, 29(18), pp. 26336–26351.

Kojima, M., Yoshida, A. and Sasaki, S. (2009) 'Difficulties in applying extended producer responsibility in developing countries', *Journal of Material Cycles and Waste Management*, 11(3), pp. 263–269.

Kükenthal, V.C., Mitchell, B., Mounzer, Y. and Rtabi, A. (2023). Optimising Recycling Policy in the UK: The UK's Deposit Return Scheme.

Kunz, N., Mayers, K. and Van Wassenhove, L.N. (2018) 'Stakeholder views on extended producer responsibility', *California Management Review*, 60(3), pp. 45–70.

Kurniawan, T.A., Avtar, R., Singh, D., Xue, W., Othman, M.H.D., Hwang, G.H., Iswanto, I., Albadarin, A.B. and Kern, A.O. (2021). 'Reforming MSWM in Sukunan (Yogjakarta, Indonesia): A case-study of applying a zero-waste approach based on circular economy paradigm', *Journal of cleaner production*, 284, p. 124775.

Leclerc, S.H. and Badami, M.G. (2020) 'Extended producer responsibility for e-waste management', *Journal of Cleaner Production*, 251, p. 119657.

Maitre-Ekern, E. (2021) 'Re-thinking producer responsibility for a sustainable circular economy', *Journal of Cleaner Production*, 286, p. 125454.

Mandpe, A., Paliya, S., Gedam, V.V., Patel, S., Tyagi, L. and Kumar, S. (2023) 'Circular economy approach for sustainable solid waste management: A developing economy perspective', *Waste Management & Research*, 41(3), pp. 499–511. https://doi.org/10.1177/0734242X221126718.

McKerlie, K., Knight, N. and Thorpe, B. (2006) 'Advancing extended producer responsibility in Canada', *Journal of Cleaner Production*, 14(6–7), pp. 616–628.

Meng, B., Qin, X. and Liu, X. (2026). 'Circular economy solutions for sustainable carbon management: Pathways to emissions reduction and a low-carbon transition', *Clean Technologies and Environmental Policy*, 28(1), p. 15.

Moh, Y. (2017) 'Solid waste management transformation in Malaysia', *Resources, Conservation and Recycling*, 116, pp. 1–14.

Möslinger, M., Ulpiani, G. and Vetters, N. (2023). 'Circular economy and waste management to empower a climate-neutral urban future', *Journal of Cleaner Production*, 421, p. 138454.

Mwanza, B.G. and Mbohwa, C. (2019) 'Strategies for enhancing extended producer responsibility enforcement', in *Proceedings of the WABER Conference*. Accra, Ghana, pp. 921–930.

Mwanza, B.G. and Telukdarie, A. (2024) 'Extended producer responsibility policies and circular economy nexus', *International Journal of Environment and Sustainable Development*, 23(4), pp. 343–366.

Parajuly, K., Fitzpatrick, C., Muldoon, O. and Kuehr, R. (2020) 'Behavioural change for sustainable e-waste management', *Journal of Cleaner Production*, 256, p. 120386.

Peng, S., Yang, Y., Li, T., Smith, T.M., Tan, G.Z. and Zhang, H.C. (2019). 'Environmental benefits of engine remanufacture in China's circular economy development', *Environmental Science & Technology*, 53(19), pp. 11294–11301.

Pinheiro, M.A.P. et al. (2019) 'New product development and circular economy', *Management Decision*, 57(4), pp. 840–862.

Puvača, N., Nikolić, A., Maletić, R. and Stević, Ž. (2025) 'Behavioural drivers of circular economy practices in the EU', *Resources, Conservation and Recycling*, 202, p. 107308.

Republic of Zambia (2004). *The Local Government Act (Cap. 281): The Lusaka (Solid Waste Management) By-laws, 2004.* Lusaka: Government Printer.

Reshetnikova, I., Lieskovska, V., Polous, O. and Murin, V. (2025) 'Deposit-return systems in circular economy', *Sustainability*, in press.

Rihm, A., Piamonte, E., Silva Villacorta, R. and Inter-American Development Bank (2024). *Digital Transformation of Solid Waste Management: Waste Collection Innovation, Business Intelligence and Smart Waste Technologies in Latin America and the Caribbean.* Washington, DC: Inter-American Development Bank. https://publications.iadb.org/en/digital-transformation-solid-waste-management-waste-collection-innovation-business-intelligence-and (Accessed: 1 July 2026).

Sharma, S. (2007). 'Exploring best practices in public–private partnership (PPP) in e-Government through select Asian case studies', *The International Information & Library Review*, 39(3–4), pp. 203–210.

Sohal, A., Nand, A.A., Goyal, P. and Bhattacharya, A. (2022). 'Developing a circular economy: An examination of SME's role in India', *Journal of Business Research*, 142, pp. 435–447.

Sosunova, I. and Porras, J. (2022) 'IoT-enabled smart waste management systems for smart cities: A systematic review', *IEEE Access*, 10, pp. 73326–73363. https://doi.org/10.1109/ACCESS.2022.3188308.

Sridan, P. and Surapolchai, S. (2020) 'Participatory action research for community-based circular waste management', *Sustainability*, 12(19), p. 8076.

Suchek, N. et al. (2021) 'Innovation and the circular economy', *Business Strategy and the Environment*, 30(8), pp. 3686–3702.

Vassileva, A. and Simić, M. (2023) 'Green public–private partnerships', *Technologies of Green Economy and Sustainable Development*, 2, pp. 161–188.

Yadav, H., Soni, U. and Kumar, G. (2023). Analysing challenges to smart waste management for a sustainable circular economy in developing countries: A fuzzy DEMATEL study', *Smart and Sustainable Built Environment*, 12(2), pp. 361–384.

Zaman, A.U. and Lehmann, S. (2011) 'Urban growth and waste management optimization towards zero-waste cities', *City, Culture and Society*, 2(4), pp. 177–187.

Zorpas, A.A. (2024) 'Transformative impact of deposit refund systems', *Waste Management & Research*, 42(12), pp. 1093–1095.

4 Sector-Specific Applications

4.1 URBAN SOLID WASTE MANAGEMENT

4.1.1 Challenges of Urban Waste Management in Developing Cities

Urban SWM continues to be a significant concern for emerging cities. Accelerated urbanization, population expansion, inadequate infrastructure, and fragile governance frameworks frequently exceed towns' capacity to handle solid waste efficiently. Consequently, numerous cities in Africa, Asia, and Latin America face escalating waste volumes, environmental deterioration, and public health hazards (Guerrero et al., 2013). This debate emphasizes the primary problems influencing urban SWM in developing countries. The following are the specific challenges for discussion in this chapter.

4.1.1.1 Rapid Urbanization and Increasing Waste Generation

Rapid urban population expansion is a primary factor contributing to urban waste management difficulties. Cities in developing nations grow more rapidly than the establishment of essential waste infrastructure and services. Oyedele (2016) observes that rapid urbanization results in garbage volumes that surpass municipal collection and disposal capabilities. Likewise, Srivastava et al. (2015) underscore that increasing living standards and shifts in consumption patterns, especially in nations such as India, are generating more varied and nonbiodegradable waste streams that are more challenging to manage.

4.1.1.2 Inadequate Waste Collection and Transportation Systems

Numerous cities exhibit inadequate garbage collection services, particularly in informal areas. Henry et al. (2006) demonstrate in their Kenyan case study that a mere portion of urban waste is collected owing to inadequate vehicles, substandard road networks, and constrained financial resources. Zohoori and Ghani (2017) note that collection efficiency is predominantly low in many low-income countries because of antiquated equipment, inadequate institutional coordination, and inconsistent service schedules.

4.1.1.3 Poor Waste Segregation and Recycling Practices

Source waste segregation is infrequent in numerous expanding urban areas. Mixed garbage hinders recycling efforts and elevates the quantity of waste that ultimately reaches landfills or open dumpsites. Abarca-Guerrero et al. (2015) emphasize that, notwithstanding the recycling potential, the informal sector frequently functions without institutional acknowledgment or assistance. Dhokhikah and Trihadiningrum (2012) also observe that Asian cities face challenges in incorporating recycling into formal municipal processes, leading to ineffective material recovery.

 DOI: 10.1201/9781003591764-4

4.1.1.4 Limited Financial Resources and Weak Institutional Capacity

Municipalities in low- and middle-income nations frequently encounter fiscal limitations that impede investment in contemporary infrastructure, recycling facilities, engineered landfills, and public awareness initiatives. Guerrero et al. (2013) contend that financial constraints are fundamental obstacles to effective SWM. Amoah and Kosoe (2014) illustrate this issue in Ghana's Wa municipality, where insufficient budget and ineffective cost recovery systems hinder efficient garbage service provision. Inadequate institutional frameworks, ambiguous duties, and disjointed agencies exacerbate the issue (Zohoori & Ghani, 2017).

4.1.1.5 Inefficient and Environmentally Unsafe Disposal Methods

Open dumping persists in numerous emerging urban areas. Srivastava et al. (2015) assert that the lack of constructed landfills, inadequate regulatory enforcement, and elevated operational costs compel towns to depend predominantly on open dumpsites, which present hazards including groundwater contamination, methane emissions, and vector-borne diseases. Henry et al. (2006) identify analogous issues in Kenya, characterized by inadequately managed disposal sites and a deficiency of environmental protections.

4.1.1.6 Low Public Awareness and Poor Community Participation

Public attitudes and behavior strongly influence waste management outcomes. Amoah and Kosoe (2014) indicate that insufficient awareness of appropriate waste management procedures and limited public engagement lead to indiscriminate dumping in urban areas. Oyedele (2016) observes that continuous community involvement is crucial for waste reduction, segregation, and adherence to anti-littering measures; however, numerous municipalities overlook initiatives aimed at behavioral change.

4.1.1.7 Governance, Policy Gaps, and Enforcement Challenges

Inadequate rules, disjointed laws, and ineffective enforcement hinder sustainable waste management initiatives. Dhokhikah and Trihadiningrum (2012) contend that governance difficulties, such as corruption, insufficient technical skills, and the lack of definitive waste management plans, hinder advancement in growing Asian cities. Abarca-Guerrero et al. (2015) also emphasize that, in the absence of robust political commitment, regulations are inadequately enforced.

4.1.2 Innovative Urban Waste Management Solutions

Urban solid waste management is transitioning from a "collect–transport–dump" approach to CE frameworks that perceive garbage as a resource and cities as centers of material circulation. Circular methods prioritize waste prevention, reuse, repair, recycling, and energy recovery above linear flows from extraction to disposal, supported by innovative technology and commercial structures. Recent studies illustrate that CE-based innovations can enhance the sustainability, climate resilience, and inclusivity of urban waste systems, particularly in developing nations (Iqbal et al., 2023; Paes et al., 2024; Izuchukwu et al., 2025).

4.1.2.1 CE Framing in Urban Waste Systems

The CE in urban areas seeks to close material loops by prolonging product lifespans, reclaiming resources from waste, and eliminating pollutants via design. Veckalne and Tambovceva (2021) underscore that advances in the CE for sustainable urban development connect waste management with urban design, energy systems, and industrial symbiosis. This indicates that MSW is regarded not merely as a liability but as a resource for new economic ventures: recycled materials, compost, bioenergy, and secondary raw materials incorporated into local value chains.

4.1.2.2 Waste Management Interventions for Climate Smart Circular Cities

Paes et al. (2024) demonstrate how focused waste management strategies in Brazilian cities can concurrently promote circularity and mitigate climate change. Illustrations encompass:

- Distinct collection of recyclables and organic waste, facilitating superior material recovery and composting.
- Assistance to recycling cooperatives, incorporating informal recyclers into official systems, and enhancing recovery rates.
- Diverting organic materials from landfills by composting or biodigestion to mitigate methane emissions.

These interventions illustrate that CE-based waste strategies encompass more than just cleaner streets; they serve as effective climate instruments that diminish greenhouse gas emissions, prolong landfill lifespan, and promote green employment in collection, sorting, and processing.

4.1.2.3 Smart Waste Management Technologies and Digital Innovation

Intelligent waste management utilizes digital technologies: sensors, IoT, data analytics, and mobile platforms to enhance urban waste distribution. Yadav et al. (2023) examine the obstacles to effective waste management for sustainable CE in underdeveloped nations through a fuzzy DEMATEL methodology. They underscore significant concerns such as:

- Inadequate digital infrastructure and technical proficiency.
- Significant initial expenditures associated with smart bins, sensors, and information and communication technology systems.
- Disjointed governance and inadequate cooperation among stakeholders.

Notwithstanding these obstacles, intelligent systems can significantly enhance circular results when executed well. Intelligent waste receptacles and sensor-integrated collection vehicles are capable of:

- Enhancing collection routes to minimize fuel consumption and emissions.
- Supervising fill levels to prevent overflow and unlawful disposal.
- Producing data to strategize infrastructure for recycling, composting, and energy recovery.

From a CE perspective, digital tools serve as enablers rather than ends; they enhance the visibility, traceability, and manageability of material flows, thus facilitating increased recycling rates and more effective resource utilization.

4.1.2.4 Waste as a Resource: Regenerative Business Models

The CE in waste management necessitates novel business strategies that regard waste streams as input resources. Iqbal et al. (2023) offer a novel regenerative CE model for Pakistan that incorporates MSW into value-enhancing operations. Principal attributes comprise:

- Material recovery facilities that categorize, cleanse, and enhance recyclables for manufacturing applications.
- Enterprises converting waste into compost, RDF, or recyclable plastics.
- Collaborations among municipalities, private enterprises, and communities to distribute risks, investments, and revenues.

These methods can enhance cost recovery for communities, provide local jobs, and diminish reliance on raw resources. Circular urban waste systems can be financially sustainable when structured as integrated value chains instead of stand-alone public services.

4.1.2.5 Smart Cities, CE, and Urban Innovation

Smart city efforts are progressively associated with CE agendas. Izuchukwu et al. (2025) investigate the advancement of waste management in Nigeria through the integration of smart city and CE strategies. They emphasize urban advances including:

- Digital platforms and mobile applications that link homeowners, waste collectors, and recyclers, facilitating door-to-door collection, on-demand services, or incentive programs (e.g., points for recyclables).
- Intelligent sorting and tracking technologies that facilitate the monitoring of plastic, metal, and organic material flows within the urban environment.
- Incorporation of waste management within comprehensive smart city frameworks, including energy, water, and transportation systems.

These developments demonstrate that CE-oriented waste management is not a stand-alone environmental strategy, but rather an integral component of a broader urban innovation ecosystem. When waste, energy, mobility, and housing are integrated in planning, communities can harness synergies, such as utilizing organic waste for biogas to fuel public buses or facilities.

4.1.2.6 Policy, Governance, and Social Inclusion in Circular Waste Systems

Technological and corporate advances require supportive policies and inclusive governance. Paes et al. (2024) underscore the significance of regulatory frameworks, climate objectives, and municipal initiatives that prioritize waste prevention and resource recovery. Veckalne and Tambovceva (2021) emphasize the necessity of integrated planning and stakeholder engagement to guarantee that circular innovations are socially endorsed and commercially viable.

Social inclusion is especially crucial in environments where informal waste collectors significantly contribute to recycling efforts. Incorporating people into formal circular systems as cooperative members, service providers, or contracted partners can augment recovery rates while enhancing livelihoods and working conditions. This corresponds with the regenerative CE paradigm advocated by Iqbal et al. (2023), wherein circularity produces both environmental and social advantages.

4.2 CASE STUDIES FROM MAJOR DEVELOPING CITIES

Urban waste management in growing countries necessitates solutions that integrate environmental sustainability, economic feasibility, and social inclusivity. Utilizing CE concepts, communities are implementing new models that convert waste into resources, promote urban symbiosis, and enhance local value chains. Two significant studies; Geng et al. (2010) and Ambati (2019), offer valuable insights into the adaptation of CE-oriented technologies inside developing country contexts. Kawasaki, Japan, while being a developed city, has inspired emerging economies with its urban symbiosis model, while Ahmedabad, India, provides a concrete illustration of CE innovation via social enterprise.

4.2.1 Urban Symbiosis Model: Lessons for Developing Economies

Geng et al. (2010) assess Kawasaki's urban symbiosis model, wherein waste from one industrial sector serves as an input for another. This paradigm is particularly pertinent for emerging cities facing challenges related to landfill saturation, inadequate recycling rates, and fragmented industrial networks. Kawasaki's CE-based approach established networks for the sharing of MSW industrial by-products and energy flows among enterprises. For instance, waste polymers are transformed into raw materials for cement kilns, while food waste is processed into animal feed or electricity.

Applying this methodology to emerging economies demonstrates significant potential. Numerous African and Asian urban areas possess concentrated industrial clusters: manufacturing zones, food processing facilities, and scrap metal markets that could gain from organized resource exchanges. By facilitating waste-to-resource transitions, local governments can diminish disposal expenses, enhance recycling rates, and earn income. The primary insight is that CE-oriented innovation necessitates not just advanced technological solutions but also interfirm collaboration, conducive legislation, and infrastructure that facilitates resource circulation.

4.2.2 Case of Ahmedabad, India: Social Innovation for Circular Waste Management

Ahmedabad shows a pragmatic instance of CE innovation integrated within community-driven frameworks in a developing economy. Ambati (2019) elucidates the transformation of urban waste management in Ahmedabad by social businesses employing inclusive, cost-effective, and scalable methods. A prominent organization collaborates with informal waste pickers, historically marginalized from formal waste management systems by providing training, organizing them into cooperatives, and incorporating them into municipal recycling frameworks.

Through these programs, waste pickers gather, categorize, and process recyclables, which are subsequently converted into secondary raw materials for local industries. Social entrepreneurs implemented decentralized composting units that transform organic waste into compost, which is subsequently sold to farmers and landscapers. This not only alleviates landfill burden but also establishes circular economic systems that assist low-income areas.

The Ahmedabad model demonstrates that CE innovation in poor nations is most effective when technology is integrated with social inclusion. The city successfully enhanced recycling rates, diminished municipal waste management costs, and created employment opportunities for marginalized populations. The strategy embodies circularity beyond materials; it revitalizes social value, fortifies local ecosystems, and embeds sustainability within community frameworks.

4.2.3 Case Study: Smart City Circular Waste Innovations in Nigeria

The swift urbanization in Nigeria has exacerbated the strain on urban waste management systems, especially in megacities like Lagos, Abuja, and Port Harcourt. Izuchukwu Precious et al. (2025) investigate the impact of integrating smart city technologies and CE practices on garbage management in several Nigerian cities. The research emphasizes various pilot projects that illustrate how urban innovation can promote circularity and sustainability.

An essential innovation is the implementation of digital waste collection platforms, enabling homes and companies to utilize mobile applications for scheduling collections, monitoring waste streams, and receiving incentives for accurate sorting and recycling. This approach enhances transparency, mitigates illegal dumping, and fortifies the connection among citizens, garbage collectors, and recycling facilities. The tools enable municipalities to analyze real-time data, facilitating effective route planning and targeted actions in high-waste areas.

A significant CE intervention is the emergence of intelligent recycling hubs including automated sorting systems. These facilities recycle plastics, metals, and electronic waste, providing secondary raw materials to regional industry. The Nigerian model effectively integrates informal waste pickers by registering them on digital platforms and connecting them to recycling hubs, hence enhancing income opportunities and system efficiency.

Lagos has implemented sensor-equipped rubbish containers that track fill levels and notify pickup teams, minimizing overflow and enhancing environmental cleanliness. These technologies, coupled with public awareness efforts, have enhanced recycling rates and diverted substantial waste from landfills. The story illustrates how smart city technologies, CE concepts, and inclusive governance may revolutionize waste management systems, especially in resource-limited environments.

4.2.4 Case Study: Circular and Community-Based Waste Innovation in Mombasa County, Kenya

Mombasa County confronts escalating waste management issues stemming from population increase, tourism demands, and insufficient disposal infrastructure. Otundo (2024) delineates a range of creative strategies facilitating the county's shift towards a circular urban waste system in accordance with the Sustainable Development Goals (SDGs).

The introduction of community-driven MRFs is one of the most significant breakthroughs. These facilities categorize plastics, paper, metals, and organic waste, reclaiming materials that would have otherwise been disposed of in landfills. Plastics are converted into pellets for production, and organic waste is composted and marketed to farmers, thus enhancing local food systems and diminishing dependence on artificial fertilizers. This circular loop endorses Sustainable Development Goals 11, 12, and 13 by advocating for sustainable urban environments, responsible production practices, and climate action initiatives.

Mombasa has implemented public–private partnerships (PPPs) to enhance circular waste solutions. Local businesses partner with the county to manage recycling facilities, gather sorted waste, and conduct educational initiatives aimed at hotels, markets, and residences. The tourism sector actively engages in beach cleanup and plastic recovery initiatives, transforming collected plastics into paver blocks and other circular products.

Another novelty is the incorporation of youth-led garbage entrepreneurs, bolstered by training and microgrants. Emerging businesses engage in plastic upcycling, compost generation, and digital waste collection services, fostering green employment and promoting behavioral transformation within communities.

The example of Mombasa demonstrates that CE strategies, when integrated with robust local engagement, enterprise assistance, and effective technologies, may significantly enhance waste management outcomes. The county's concept exemplifies a scalable framework for other emerging urban regions pursuing affordable, inclusive CE solutions.

4.3 RURAL SOLID WASTE MANAGEMENT

Management of rural solid waste in emerging nations is widely acknowledged as an essential element of sustainable development; nonetheless, it remains much under-explored and underfunded in comparison with urban waste systems. Rural regions produce reduced waste quantities; nonetheless, inadequate infrastructure, constrained institutional capability, and minimal public awareness lead to extensive dumping, open incineration, and environmental deterioration. From a CE viewpoint, these difficulties simultaneously offer potential to establish cost-effective, community-oriented waste systems that emphasize material recovery, local value generation, and environmental restoration. Vinti and Vaccari (2022) emphasize that rural communities frequently lack formal garbage collection services, resulting in extensive contamination; however, they also exhibit robust social cohesiveness and local knowledge that can facilitate inclusive, circular solutions. CE strategies such as decentralized composting, small-scale recycling centers, and waste-to-resource microenterprises can convert rural waste streams into valuable resources for agriculture and local industry.

Plastic pollution poses significant challenges in rural areas, where inadequate disposal facilities and the increasing availability of packaged products generate ongoing environmental hazards. Mihai et al. (2021) observe that plastics accumulate in streams, agricultural lands, and public areas owing to indiscriminate dumping, jeopardizing soil fertility, animal health, and rural economies. Nevertheless, the authors also recognize circular opportunities: rural regions can implement low-tech plastic recycling initiatives, advocate for reusable packaging methods, and incorporate informal waste collectors into community recycling programs. These projects not only mitigate pollution but also create local revenue and bolster rural economies. In numerous instances, women and adolescents have been pivotal to these initiatives, illustrating how CE methodologies may facilitate social inclusion and empowerment.

Mihai (2023) emphasizes the connection between CE tactics and rural development, asserting that CE principles inherently correspond with rural sustainability objectives, such as resource efficiency, regenerative agriculture, and local entrepreneurship. Rural communities currently engage in circular practices such as reuse, repair, and organic waste valorization; formalizing and expanding these practices allows rural areas to mitigate environmental consequences while enhancing economic resilience. Opportunities encompass converting agricultural leftovers into compost or bioenergy, creating community repair centers, and incorporating circular ideas into rural tourist and craft industries. These strategies enhance local value chains and diminish rural reliance on external resources.

Implementing CE-based waste management in rural regions necessitates overcoming many obstacles, as demonstrated by Rimantho et al. (2023) in their research on Jepara District, Indonesia. Principal challenges encompass insufficient finance, restricted technical proficiency, absence of policy endorsement, feeble enforcement, and minimal community engagement. The research proposes measures including capacity enhancement for local leaders, establishment of village-level waste banks, incentives for home waste separation, and collaborations with the business sector to bolster small-scale recycling enterprises. These approaches can assist rural governments and communities in transitioning from linear waste practices to circular, sustainable systems.

A CE concept provides rural communities in emerging economies with a means to minimize waste, create livelihoods, and safeguard ecosystems. Through the integration of local expertise, decentralized technologies, community engagement, and supportive governance frameworks, rural regions may convert garbage from a continual obstacle into a driver of sustainable development and economic potential.

4.3.1 Unique Challenges of Rural Waste Management

Rural SWM in developing economies faces unique issues due to scattered populations, constrained financial resources, and inadequate institutional support. In contrast to metropolitan systems, rural areas frequently lack institutional waste management infrastructure, although they are progressively producing varied and nonbiodegradable waste streams. The subsequent subsections delineate the principal issues highlighted in the current study.

4.3.1.1 Limited Waste Collection Infrastructure and Services

A fundamental difficulty in rural regions is the lack of organized garbage collection and disposal systems. Vinti and Vaccari (2022) emphasize that most rural settlements in developing nations are devoid of formal municipal services, leading to prevalent open dumping and incineration. Scattered settlement patterns render waste collection logistically challenging and economically unappealing for municipal agencies. Wang et al. (2017) similarly claim that numerous villages in rural China lack fundamental waste management infrastructure, resulting in environmental contamination of agricultural lands and waterways.

4.3.1.2 Financial Constraints and Inadequate Municipal Budgets

Rural communities sometimes function with constrained financial resources, limiting their capacity to invest in waste infrastructure, manpower, and technology. Mihai (2017) elucidates that rural waste management is often inadequately funded due to national budgets and donor resources predominantly favoring urban regions. Moharana (2014) notes that income-collecting mechanisms in rural communities are either inadequate or absent, hindering the sustainability of waste services and the implementation of enhanced systems.

4.3.1.3 Low Public Awareness and Cultural Practices

A significant number of rural families are uninformed about appropriate waste disposal and the implications for environmental health. Conventional methods including open burning, indiscriminate waste disposal, and dumping in agricultural zones or watercourses persist (Moharana, 2014). Vinti and Vaccari (2022) observe that behavioral modification is sluggish in rural environments due to entrenched cultural attitudes, insufficient knowledge regarding waste matters, and minimal exposure to awareness initiatives. Boateng et al. (2023) highlight analogous difficulties in rural Ghanaian schools, where insufficient understanding leads to ineffective waste segregation and hazardous disposal practices.

4.3.1.4 Increasing Volumes of Nonbiodegradable Waste

Contemporary consumption trends are importing plastics, packaging, and technological garbage into rural environments. Mihai (2017) underscores that the composition of rural waste has undergone considerable transformation, although systems continue to be tailored for organic agricultural waste. Wang et al. (2017) caution that rural China is increasingly plagued by the contamination of soils and water supplies due to plastic bags, bottles, and single-use products. The discord between conventional waste management procedures and emerging waste categories poses environmental and health hazards.

4.3.1.5 Lack of Skilled Personnel and Institutional Capacity

Rural authorities frequently lack qualified personnel in waste management, environmental health, or CE methodologies. Sharma et al. (2025) observe that constrained human resources result in inadequate planning, monitoring, and enforcement. Numerous rural regions also lack the technical proficiency necessary to implement novel solutions such as composting technologies, recycling micro-centers, or circular farming models. Consequently, even meticulously crafted policies encounter difficulties in achieving success.

4.3.1.6 Weak Policy Enforcement and Fragmented Governance

Regulatory frameworks for rural solid waste management are sometimes ambiguous or inadequately enforced. Vinti and Vaccari (2022) assert that governance in rural waste systems is often fragmented, characterized by the operation of many entities lacking coordination. Local councils may lack the authority or capacity to enforce regulations. Wang et al. (2017) demonstrate that in rural China, despite national regulations advocating for waste enhancements, enforcement at the village level is inadequate.

4.3.1.7 Environmental and Public Health Impacts

Inadequate garbage disposal results in the contamination of surface water, groundwater, and agricultural soils. Boateng et al. (2023) discovered that rural schools encounter heightened littering, unpleasant odors, vector infestations, and contamination hazards. Mihai (2017) underscores that these environmental effects are frequently more pronounced in rural communities due to the intimate interconnection among households, agricultural fields, and natural ecosystems.

4.3.1.8 Limited Integration of CE Practices

Despite the considerable potential for circular waste solutions in rural areas such as composting, reusing, and recycling, implementation progresses at a sluggish pace. Sharma et al. (2025) identify limitations such as insufficient training, little understanding of CE benefits, minimal private sector involvement, and inadequate incentives for households or community organizations. Vinti and Vaccari (2022) also observe that rural communities are not completely capitalizing on potential to transform organic waste into agricultural resources or to start small-scale recycling firms.

4.3.2 Tailored Solutions for Rural Areas

Developing efficient SWM systems for rural regions necessitates solutions that acknowledge sparse population density, constrained financial resources, robust community frameworks, and elevated proportions of organic waste. The literature emphasizes a collection of context-specific, frequently low-cost interventions that can be far more effective than merely replicating metropolitan models.

4.3.2.1 Decentralized and Low-Cost Waste Systems

Rural communities seldom validate costly, centralized infrastructure. Vinti and Vaccari (2022) underscore decentralized SWM systems centered on village-level solutions, including tiny transfer stations, straightforward collecting methods, and locally administered disposal or treatment facilities. These technologies minimize transportation distances and expenses while enabling communities to tailor the infrastructure to their own location and requirements. Basic alternatives like community collection sites, communal tractors or carts, and locally constructed composting pits are frequently more feasible than comprehensive truck-based systems intended for urban areas.

4.3.2.2 Community-Based and Cooperative Approaches

Due to the typically greater social cohesion in rural regions, community-based management can serve as an effective paradigm. Sharma et al. (2025) emphasize the efficacy of village committees, women's groups, youth clubs, and cooperatives in administering collection services, overseeing composting units, or operating small recycling centers. These groups can facilitate door-to-door awareness campaigns, fee collection, and local enforcement of regulations. When communities possess a sense of ownership over the system, they are more inclined to segregate waste, remit reasonable user fees, and safeguard communal amenities from vandalism.

4.3.2.3 CE Strategies for Organic and Plastic Waste

Rural waste predominantly consists of organic materials and plastics, both of which provide significant CE) prospects. Mihai et al. (2021) emphasize that organic waste can be valorized through:

- Household or community composting that bolsters local agriculture.
- Compact biogas systems for culinary energy or illumination.
- Utilization of agricultural wastes for mulching or soil enhancement.

Concurrently, plastics can be managed with low-tech circular solutions such as:

- Rural "waste banks" where households trade sorted plastics for currency or commodities.
- Collaborations with SMEs that shred and pelletize plastics for utilization in local manufacturing.
- Advocacy for reusable packaging and refill systems via rural retail outlets and cooperatives.

These CE strategies mitigate pollution, enhance local livelihoods, and maintain value circulation within rural economies.

4.3.2.4 Smart Planning with Remote Sensing and GIS

Digital tools can markedly enhance efficiency even in remote environments. Cheng et al. (2024) demonstrate that high-resolution remote sensing and GIS can be utilized to:

- Chart dispersed residences, thoroughfares, and unofficial waste disposal sites.
- Revise efficient collection routes for a constrained number of vehicles.
- Determine optimal placements for community bins, transfer stations, or treatment facilities.

Through spatial analysis, rural authorities can optimize fuel expenditures, decrease journey duration, and enhance service coverage, even with the availability of only one or two cars. These instruments are particularly advantageous in extensive districts characterized by challenging topography and scattered settlements.

4.3.2.5 Integrated Education and Behavior Change

All four studies emphasize that technical solutions alone are insufficient. Customized solutions must incorporate ongoing education and behavioral modification programs. Vinti and Vaccari (2022) and Sharma et al. (2025) advocate for school-based initiatives, demonstration projects (such as conspicuous composting sites), and peer-to-peer education via farmer groups or cooperatives. When households comprehend the connection between waste, soil health, water quality, and livelihoods, they are more inclined to implement segregation, refrain from incineration, and engage in circular projects.

4.3.3 Successful Rural Waste Management Programs

Effective rural waste management initiatives illustrate that, even in low-income, resource-limited environments, communities can develop sustainable, inclusive systems that substantially diminish pollution and create new economic prospects. Leknoi et al. (2024) present a comprehensive analysis of a rural waste management program that converted a community into a paradigm of environmental stewardship and CE(CE) innovation. Their findings provide significant insights for other developing regions aiming to enhance rural waste management techniques.

The rural initiative reported by Leknoi et al. (2024) was founded on community empowerment, robust local leadership, and pragmatic, cost-effective technologies. The effort centered on the formation of a community-driven waste management committee tasked with the planning, implementation, and oversight of local waste systems. This grassroots approach guaranteed that strategies were consistent with communities' social norms, economic conditions, and environmental requirements. The committee engaged households via frequent conversations, village meetings, and educational initiatives, progressively enhancing public participation and altering community behaviors away from open dumping and incineration.

A key strategy of the program was the implementation of systematic household waste segregation, focusing on the division of organic materials, recyclables, and residual waste. Residents received training on proper garbage separation and were supplied with basic sorting containers constructed from locally sourced materials. Leknoi et al. (2024) observe that this behavioral modification substantially augmented the quantity of recyclable items collected, concurrently diminishing environmental contamination and cutting overall waste management expenses.

To facilitate the shift to a circular economy, the community developed local garbage processing centers. Organic waste was directed into decentralized composting facilities, generating nutrient-dense fertilizer utilized by farmers and household gardeners. This closed-loop methodology diminished dependence on chemical fertilizers, decreased household expenditures, and enhanced soil health, illustrating how rural waste management can directly bolster agricultural yield. Recyclables, including plastics, glass, and metals, were sold to local recycling companies, producing revenue that the community reinvested into environmental projects.

A notable characteristic of the program was its dependence on inclusive social innovation. Women's organizations, youth associations, and school clubs participated actively in garbage segregation, awareness initiatives, and innovative upcycling endeavors. These groups manufactured bags, domestic items, and crafts using low-value recycled materials, establishing new micro-enterprises. Leknoi et al. (2024) emphasize that this inclusive strategy improved community cohesion, bolstered local livelihoods, and elevated the prominence of environmental stewardship in the village.

The program's success was additionally bolstered by collaborations with local governmental and nongovernmental organizations. These entities offered technical training, logistical assistance, and little financial support that facilitated the establishment of waste storage facilities and composting units. The initiative highlighted the significance of multi-stakeholder collaboration, wherein local knowledge is enhanced by external experience to provide sustainable and scalable solutions.

Leknoi et al. (2024) assert that the rural waste program was successful due to its incorporation of behavior modification, CE principles, community stewardship, and social entrepreneurship. It demonstrates how rural communities frequently neglected national waste strategies can lead the way in sustainable waste models that safeguard the environment, bolster local economies, and encourage broader regional implementation. The scenario offers a repeatable framework for developing nations aiming to convert rural waste issues into prospects for circular, community-focused development.

4.4 INDUSTRIAL SOLID WASTE MANAGEMENT

Industrial solid waste management is increasingly incorporating CE principles that prioritize resource efficiency, waste reduction, and material recovery during production processes. In developing countries, where industrial waste often intensifies environmental degradation, the CE model provides a method to transform waste into valuable resources while promoting sustainable economic growth. Ezeudu and Ezeudu (2019) assert that Nigerian enterprises have initiated the adoption of CE strategies, such as material substitution, closed-loop recycling, and industrial symbiosis, allowing manufacturing sectors to reduce waste generation and generate secondary raw materials. These methods not only diminish disposal costs but also create new economic opportunities from recovered metals, polymers, solvents, and biomass by-products.

Awuchi et al. (2020) assert that industries employing circular economy-driven waste management exhibit enhanced environmental performance, reduced greenhouse gas emissions, and increased global competitiveness. Many countries are enacting legal frameworks and sustainability mandates that require businesses to minimize waste generation and adopt life cycle analysis. This transition is evident in China's industrial sector, where CE initiatives have been integrated into national strategies. Tang et al. (2020) illustrate that China's circular industrial parks and resource-sharing networks have led to measurable improvements in waste reduction and material recovery, demonstrating how coordinated spatial design and industry collaboration may accelerate the advantages of a circular economy.

Naqvi et al. (2021) contend that effective industrial waste management within a CE requires technological innovation, reorganization of production processes, and collaboration across value chains. They note that cleaner manufacturing techniques, source waste segregation, and product design for durability and recyclability substantially reduce industrial waste loads. CE-oriented industrial waste management facilitates a shift from linear "take–make–dispose" systems to regenerative ones that recycle materials, conserve resources, and encourage sustainable economic development. This transition is essential for developing economies seeking to harmonize economic expansion with environmental conservation.

4.4.1 Handling Industrial and Hazardous Waste

Effective management of industrial and hazardous waste is essential for safeguarding public health, protecting ecosystems, and promoting sustainable industrial development. Hazardous waste, comprising chemicals, solvents, heavy metals, sludge, and contaminated residues, requires management methods to prevent soil, air, and water pollution. Nemerow and Agardy (1998) contend that efficient hazardous waste management must include prevention, minimization, treatment, and safe disposal. Their framework underscores the importance of altering processes, substituting materials, and adopting cleaner manufacturing techniques to reduce hazardous waste generation at the source before resorting to more complex solutions.

In industrial settings, the complexity of hazardous materials requires a methodical and risk-based approach. Lindgren (2018) asserts that organizations must implement comprehensive waste management strategies that include waste identification, classification, proper storage, secure transportation, and monitoring of treatment processes. These plans must comply with regulatory criteria and incorporate continuous training for those handling hazardous products. Source segregation is crucial, as the combination of hazardous and nonhazardous waste increases treatment costs and heightens environmental risks.

Tskhovrebov et al. (2019) assert that the planning for the ecologically safe handling of highly hazardous waste requires robust infrastructure, contingency protocols, and preparedness for emergencies. Their research in the industrial, construction, and transportation sectors demonstrates that risk assessment, technological improvements, and containment measures are essential for averting accidental releases. The authors assert that businesses must use modern treatment technologies such as stabilization, incineration, and secure landfilling to ensure the safe and permanent disposal of hazardous waste.

Insights from Chile, as expressed by Navia and Bezama (2008), demonstrate the challenges faced by developing nations in the handling of hazardous industrial waste. Inadequate treatment facilities, feeble regulatory enforcement, and dependence on outdated technologies have historically impeded progress. Their research illustrates that enhancing environmental regulations, promoting private sector investment in treatment facilities, and optimizing industrial monitoring systems significantly elevate national hazardous waste management proficiency.

Efficient management of industrial and hazardous waste requires a thorough system that prioritizes prevention, utilizes advanced treatment technologies, and has strong institutional and regulatory frameworks. These approaches reduce environmental risks, improve workplace safety, and foster sustainable industrial progress.

4.4.2 CE Practices for Industrial Waste

Principles of the CE are revolutionizing industrial waste management by enhancing resource efficiency, preventing waste, and regenerating materials within production cycles. Instead of viewing waste as an inevitable by-product, the CE(CE) reinterprets it as a valuable resource that can be reprocessed through technological, operational, and organizational advances. Literature suggests that enterprises worldwide, particularly in developing nations, are rapidly adopting CE models to mitigate environmental concerns and enhance economic competitiveness.

4.4.2.1 Waste Minimization Through Cleaner Production and Process Redesign

A fundamental CE technique in industrial operations is minimizing waste at the source via cleaner production and process reengineering. Nemerow (2010) emphasizes the necessity of modernizing industrial processes, implementing precise technology, and substituting hazardous materials with safer alternatives to mitigate waste generation. Adam et al. (2025) affirm this approach, observing that sectors within industrial chemistry are progressively embracing green chemistry principles, energy-efficient systems, and process optimization to minimize by-products and enhance resource use. These solutions provide environmental advantages and financial savings.

4.4.2.2 Resource Recovery and Material Looping

CE promotes recovery, reuse, and recycling of materials to complete resource cycles. Ezeudu and Ezeudu (2019) examine industrial sectors in Nigeria and document successful instances of material recovery systems that convert metal scraps, plastic pieces, and biomass leftovers into secondary raw materials for local production. Tawo (2025) further underscores waste-to-resource solutions, including industrial symbiosis, when waste outputs from one facility serve as inputs for another. These techniques diminish dependence on raw commodities and bolster local circular value chains.

4.4.2.3 Circular Industrial Engineering and Technological Innovation

Industrial engineering is crucial in promoting the use of CE practices. Nwamekwe and Chikwendu (2025) contend that CE techniques in engineering encompass product designing for durability, modularity, and disassembly; implementation of automation for precise waste sorting; and the introduction of improved recycling technologies. These advancements allow industry to prolong product longevity and improve recoverability. Emerging technologies such as digital twins and IoT-based monitoring also support more efficient waste tracking and resource optimization.

4.4.2.4 Knowledge, Attitudes, and Organizational Culture for CE Adoption

The effective execution of CE techniques relies on both technology and the organizational culture, as well as employee understanding. In their 2025 study on industrial waste management in China, Ling, Jaafar, and Ismail assert that employee knowledge, attitudes, and behaviors significantly impact CE outcomes. Industries that implement proactive training and CE literacy initiatives, and demonstrate leadership commitment exhibit greater adoption of circular models. This underscores the necessity for capacity enhancement and behavioral transformation across all organizational tiers.

4.4.2.5 Circular Supply Chains and Collaborative Networks

The use of CE principles transcends individual enterprises, including whole supply chains. Elia et al. (2020) illustrate that sectors adopting circular procurement, reverse logistics, and collaborative waste-processing networks attain markedly superior circularity performance. Collaboration with suppliers, recyclers, and downstream partners guarantees the reintegration of recovered materials into production cycles, thereby enhancing industrial resilience and augmenting environmental performance.

4.4.2.6 Policy Support and Sustainable Operational Strategies

Effective waste management solutions are crucial for facilitating CE in industrial environments. Tawo (2025) asserts that industries necessitate facilitating policies such as extended producer responsibility (EPR), waste reduction objectives, and incentives for recycling investments to expand circular models. Incorporating CE principles into operational strategies enables companies to synchronize environmental objectives with production planning, compliance frameworks, and sustainable development trajectories.

4.4.3 Case Studies of Industrial Waste Management

Industrial waste management continues to pose a significant challenge in numerous emerging nations, as swift industrialization sometimes surpasses environmental laws, technical capability, and institutional preparedness. Numerous case studies from Lebanon, Nigeria, and Southern Africa illustrate the challenges and rising prospects for incorporating sustainable and circular practices into industrial waste systems.

4.4.3.1 Industrial Waste Management in Lebanon

El-Fadel et al. (2001) present a thorough evaluation of Lebanon's industrial waste situation amid a phase of economic growth and inadequate regulatory oversight. Their research indicates that industrial waste was primarily handled by informal, uncontrolled disposal methods, such as direct discharge into rivers, open land dumping, and co-disposal with municipal garbage. Primary challenges encompassed erratic waste data, insufficient monitoring systems, and deficient treatment facilities. Hazardous wastes, including solvents, heavy metals, and chemical residues, posed significant challenges due to the lack of specialized processing and disposal facilities.

Notwithstanding these limitations, the report demonstrates initial efforts to enhance waste management via regulatory reforms, stakeholder engagement, and pilot initiatives for waste segregation and pretreatment. The Lebanese situation illustrates the necessity for cohesive national strategy, regulatory transparency, and investment in treatment technologies to shift from sporadic disposal to sustainable industrial waste management.

4.4.3.2 CE-Based Industrial Waste Practices in Nigeria

Ezeudu and Ezeudu (2019) emphasize effective implementations of CE principles in several Nigerian businesses, demonstrating how resource recovery tactics can transform waste into economic prospects. Their case studies encompass metal fabrication industries, plastic recycling businesses, and agro-industrial processors. In these sectors, waste materials, including metal scraps, plastic offcuts, and biomass residues, were salvaged, recycled, and reintegrated into production cycles, therefore diminishing environmental impacts and raw material expenses. Industrial symbiosis networks were recorded, wherein waste products from one industry functioned as inputs for another. These activities were facilitated by entrepreneurial innovation, the expansion of recycling markets, and heightened awareness of the benefits of the circular economy. The Nigerian stories illustrate that, even in resource-limited environments, firms may implement circular models that create employment, enhance resource efficiency, and minimize landfilling.

4.4.3.3 Decision Support Framework for the Iron and Steel Industry in Southern Africa

Schoeman et al. (2021) examine industrial waste management within the iron and steel sector of Southern Africa, a waste-intensive business that produces slag, dust, sludge, and many hazardous byproducts. Their research introduces a decision support system aimed at assisting industrial managers in assessing waste treatment alternatives according to environmental, economic, and technological parameters. The framework was evaluated in a prominent steel manufacturing facility, where it facilitated the identification of potential for recycling steel slag into construction materials, recovering metals from dust particulates, and enhancing on-site waste segregation. The instance demonstrates that organized decision-making tools can direct enterprises towards sustainable waste solutions, particularly in managing complicated waste streams. The approach underscores the significance of life cycle evaluation, regulatory adherence, and intersectoral coordination.

4.4.3.4 Synthesis and Lessons

Several similar themes emerge across these case studies. Developing nations frequently contend with insufficient infrastructure and regulatory frameworks, which impede safe and efficient garbage management. Second, CE strategies such as material recovery, recycling, and industrial symbiosis provide promising avenues to mitigate environmental externalities while augmenting economic production. Ultimately, decision support instruments and organized management frameworks are essential in directing enterprises towards sustainable practices, particularly in sectors characterized by intricate waste profiles.

4.5 AGRICULTURAL AND ORGANIC WASTE MANAGEMENT

The management of agricultural and organic waste is pivotal to promoting a CE, particularly in areas where agriculture and food production yield substantial quantities of biodegradable waste. The CE concept regards organic waste not as a disposal issue but as a valuable bioresource that can facilitate soil regeneration, renewable energy generation, and sustainable food systems. Masullo (2017) contends that a circular strategy reestablishes the connection between urban and rural regions by transforming organic waste such as food remnants, manure, and agricultural by-products into compost, biogas, and nutrient-dense fertilizers that reintegrate into agricultural output. This mitigates landfill utilization, diminishes greenhouse gas emissions, and lessens reliance on synthetic inputs.

Kaszycki et al. (2021) demonstrate the integration of organic waste treatment with wastewater management in Poland's bio-based circular systems to generate high-quality bioproducts. Municipalities and farms utilize anaerobic digestion and fermentation technology to produce biogas, biofertilizers, and organic amendments, exemplifying the potential of closed-loop nutrient cycles. Their research demonstrates that implementing CE frameworks can enhance waste treatment efficiency, generate renewable energy, and augment soil quality.

In the agricultural sector, CE plans encompass not only waste management but also integrated farming systems. Singh et al. (2022) assert that agricultural residues such as crop stalks, husks, leaves, and animal manure can be converted into compost, mulch, biochar, and biochemicals. These measures improve soil structure, augment water retention, and promote sustained agricultural yield. Additionally, on-farm composting and vermicomposting provide economical, decentralized approaches for the valorization of organic waste, especially in rural areas.

Selvan et al. (2023) enhance this perspective by demonstrating how integrated organic farming techniques bolster food security and promote rural sustainability. In these systems, waste streams from agriculture, livestock, and agro-processing are perpetually recycled: manure is transformed into compost, crop remnants nourish livestock, and biogas systems supply clean energy for houses and farms. This circular integration diminishes external inputs, bolsters resistance, and increases the overall ecological equilibrium of agricultural ecosystems.

In conclusion, circular economy-focused agricultural and organic waste management promotes nutrient cycling, renewable energy production, and sustainable agricultural systems. Transforming garbage into valuable bioresources fosters climate mitigation, enhances soil health, supports rural livelihoods, and promotes long-term sustainability.

4.5.1 Importance of Managing Agricultural Waste

Efficient management of agricultural waste is crucial for environmental protection, soil fertility enhancement, public health improvement, and the promotion of sustainable agricultural systems. As agricultural practices proliferate to satisfy the food requirements of increasing populations, they produce substantial amounts of residues such as crop stalks, husks, manure, and processing by-products that can either pose environmental challenges or serve as useful resources, contingent upon their management. Maji et al. (2020) assert that the incorrect disposal of agricultural waste, such as incineration and unregulated dumping, substantially adds to air pollution, greenhouse gas emissions, soil deterioration, and water contamination. The repercussions immediately influence rural populations and ecosystems, rendering agricultural waste management an essential aspect of environmental preservation.

Xue et al. (2016) emphasize that agricultural waste constitutes a significant portion of global waste streams, highlighting the critical role of agriculture in comprehensive waste management initiatives. They observe that unregulated leftovers can obstruct streams, harm natural habitats, and diminish biodiversity. Conversely, when effectively managed, agricultural waste can be transformed into compost, bioenergy, biofertilizers, and animal feed, thereby promoting circular resource flows and diminishing reliance on synthetic inputs.

Stentiford et al. (2004) emphasize the global significance of agricultural waste management, asserting that agriculture is pivotal in sustainable waste strategies globally. They assert that incorporating agricultural waste management into national waste policy can markedly alleviate environmental burdens while improving food safety and soil quality. Such strategies are particularly vital in developing nations, where agriculture continues to be a primary source of subsistence.

Decomposers play a crucial part in the processing of agricultural waste. Iqbal et al. (2020) assert that microbes and soil fauna are essential agents in the decomposition of organic leftovers, facilitating the natural recycling of nutrients into the soil. Their function facilitates composting systems and integrated agricultural models that augment soil structure and elevate crop yield.

Effectively managing agricultural waste is essential for environmental sustainability, resource efficiency, and resilient food systems, rendering it a critical concern for contemporary agriculture.

4.5.2 Circular Practices for Agricultural Waste

Organic waste, including food remnants, agricultural by-products, garden refuse, and biodegradable industrial materials, constitutes a significant worldwide waste stream and poses both a crucial challenge and an opportunity within the CE. In conventional linear systems, organic waste is frequently disposed of in landfills or by open dumping, resulting in methane emissions, nutrient depletion, and environmental deterioration. The CE paradigm reinterprets organic waste as a regenerative resource that may be perpetually reintegrated into ecological and economic systems. Organic waste can generate value through composting, anaerobic digestion, soil application, and integrated farming methods by providing renewable energy, enhancing soil fertility, decreasing reliance on synthetic inputs, and fortifying sustainable food systems.

Recent studies emphasize the innovations in cities, rural areas, and industrial systems aimed at closing biological nutrient loops. Urban circular systems, exemplified by those in Amsterdam, integrate decentralized collection, digital monitoring, and diverse valorization of organic waste to establish more robust and low-emission waste infrastructures. Rural and agricultural settings illustrate how integrated organic farming and bio-based waste management can provide zero-waste systems that support food security and climate resilience objectives.

Circular organic waste management can diminish greenhouse gas emissions, foster renewable energy, enhance regenerative agriculture, and generate new economic opportunities in both high-income and low-income environments. However, realizing this promise necessitates robust governance, legislative coherence, technology advancement, and civic engagement. The subsequent sections demonstrate how integrating viewpoints from Europe, Asia, and developing nations may convert organic waste from an environmental burden into a fundamental resource for sustainable development through circular practices.

4.5.2.1 Circular Organic Waste Management Through Urban Systems and Design

Circular approaches for organic waste in urban environments emphasize the reconfiguration of waste streams to ensure that biological elements perpetually reintegrate into the economy. Viva et al. (2020) underscore Amsterdam as a prime exemplar of urban transformation from linear disposal paradigms to integrated circular systems. Their research indicates that successful organic waste management commences with meticulous source separation, facilitating the clean collection of food scraps and biodegradable materials for targeted valorization processes. Essential design elements comprise decentralized collecting hubs, community-level composting facilities, and logistics systems that reduce transportation impacts.

In Amsterdam's circular concept, organic waste is channeled into several value creation paths. Composting facilities generate soil amendments for urban agriculture and landscaping, whilst anaerobic digestion plants produce biogas that is utilized in district heating networks or refined for vehicular fuel. Viva et al. (2020) assert that digital tracking instruments and stakeholder engagement platforms enhance system efficiency through the promotion of transparency, real-time monitoring, and community involvement.

This approach emphasizes that circular management of organic waste is not merely a technical issue but also a socio-institutional one, necessitating the synchronization of municipal policy, private sector innovation, and resident behavior. Masullo (2017) advocates for cities to reconceptualize organic waste as a connection between urban and rural regions. When adequately handled, urban organic waste transforms into a nutrient stream capable of replenishing soils in peri-urban and rural areas. This urban–rural connection enhances circular nutrient cycles and diminishes reliance on synthetic fertilizers.

Urban circular practices seek to establish regenerative systems that consistently valorize organic waste, minimize emissions, and engage residents in maintaining circular flows.

4.5.2.2 Regulating Application of Organic Waste to Soils

The application of organic waste to soils is a well-established cyclic technique that converts biodegradable leftovers into beneficial soil nutrients. Arias et al. (2022) investigate the effects of compost, biosolids, and processed organic residues on soil structure, water retention, and nutrient availability when applied correctly. These circular solutions enhance agricultural output while concurrently redirecting substantial amounts of organic waste from landfills, where it would otherwise produce methane.

Legislative complexities frequently influence the viability of soil application tactics. Arias et al. (2022) emphasize that regulatory frameworks differ significantly among nations and frequently fail to adapt to the advancing CE models. Rigorous regulations on pollutant thresholds, heavy metals, and pathogen levels safeguard environmental integrity but may provide obstacles for municipalities or small-scale garbage processors. Aligning legislation to support CE objectives while upholding safety requirements is essential.

Haque et al. (2023) elucidate that valorizing organic waste in agriculture yields significant economic and ecological advantages. Agricultural waste streams, including crop residues, manure, and agro-processing by-products, are reintegrated into the soil via composting, fermentation, or biochar formation, thereby closing nutrient loops. This diminishes dependence on synthetic fertilizers, alleviates soil deterioration, and promotes climate-resilient agriculture. Moreover, these circular channels establish novel business forms like compost cooperatives, rural biorefineries, and farmer-led input firms.

Kaszycki et al. (2021) enhance the discourse by introducing Poland's bio-based CE model, wherein organic waste and wastewater sludge are co-processed to provide high-quality soil amendments. Their cohesive strategy illustrates how the amalgamation of technical advancement, regulatory enhancements, and stakeholder involvement can fully realize the potential of organic waste in CE systems.

Therefore, the application of organic waste to soil is fundamental to circularity, concurrently enhancing agriculture, minimizing waste, and aiding climate mitigation.

4.5.2.3 Bioenergy Recovery Through AD and Biogas Production

Bioenergy recovery is a fundamental aspect of circular organic waste management. The conversion of organic waste into biogas by AD diverts garbage from landfills while generating sustainable energy and nutrient-dense digestate. Ellacuriaga et al. (2021) emphasize the adaptability of anaerobic digestion in managing various organic waste streams, such as food waste, manure, sewage sludge, and agricultural residues, while generating biogas suitable for energy, heat, or conversion to biomethane.

The cyclic value of AD is in its capacity to produce numerous things concurrently. In addition to energy, the digestate functions as a premium fertilizer, facilitating nutrient cycles among food intake, waste production, and agricultural output. Ellacuriaga et al. (2021) demonstrate that the incorporation of anaerobic digestion into regional waste systems enhances energy security, diminishes greenhouse gas emissions, and reduces reliance on fossil-based fertilizers.

Kaszycki et al. (2021) offer another viewpoint via their Polish case study, illustrating how the integration of anaerobic digestion with wastewater treatment improves system efficiency and resource recovery. These integrated bio-based systems adhere to CE principles by optimizing the value derived from organic waste while reducing pollution.

Successful bioenergy recovery necessitates supportive policy, technical expertise, and investment. Ddiba et al. (2022) observe that in low- and middle-income nations, infrastructure deficiencies, elevated capital expenditures, and restricted technical proficiency frequently hinder adoption. Nonetheless, community-scale digesters and modular anaerobic digestion technologies provide viable alternatives for rural and peri-urban settings.

Bioenergy recovery is a significant circular technique that converts organic waste into renewable energy and agricultural inputs, enhancing energy sustainability, climate resilience, and regenerative farming systems.

4.5.2.4 Circular Agriculture and Zero-Waste Farming Systems

Circular methodologies for organic waste are fundamentally integrated into sustainable agriculture frameworks. Kotyal (2023) asserts that circular agriculture fosters zero-waste farming by reintegrating all organic outputs, crop wastes, manure, and food processing by-products into production cycles. This may manifest as composting, mulching, integrated livestock-crop systems, agroforestry, and charcoal production. These routes augment soil fertility, mitigate nutrient losses, and bolster farm resilience.

Selvan et al. (2023) observe that integrated organic farming establishes multiple circular loops: animal dung fertilizes crops, crop residues nourish cattle, and organic waste powers biogas units. These interconnected systems diminish external input expenses while facilitating climate adaptability. Closed-loop nutrient flows mitigate environmental pollution by minimizing runoff, soil deterioration, and methane emissions linked to unregulated agricultural waste.

Masullo (2017) asserts that circular agriculture reestablishes the connection between rural and urban systems by facilitating the return of organic waste from cities to farms in the form of compost or digestate. This urban–rural nutrient cycle enhances food security, alleviates strain on landfill systems, and bolsters regional sustainability.

Haque et al. (2023) underscore that agricultural industries derive substantial advantages from implementing organic circular waste techniques, particularly in underdeveloped nations where soil deterioration and input affordability present considerable obstacles. Transforming waste into value benefits smallholder farmers, local businesses, and rural economies.

Circular agriculture exemplifies the utilization of organic waste as a regenerative resource vital to sustainable agricultural systems, enhancing soil health, biodiversity, and climate resilience.

REFERENCES

Abarca-Guerrero, L., Maas, G. and Hogland, W. (2015) 'Solid waste management challenges for cities in developing countries', *Revista Tecnología en Marcha*, 28(2), pp. 141–168.

Adam, A.B., Abubakar, M.Y. and Filibus, D. (2025) 'Circular economy in industrial chemistry: A review of strategies for waste reduction and resource recovery', *Chemical Research and Technology*, 2(3), pp. 140–145.

Ambati, N.R. (2019) 'Social innovation practices in sustainable waste management: Case study of successful social enterprises in Ahmedabad', *International Journal of Scientific and Technology Research*, 8(12), pp. 1978–1985.

Amoah, S.T. and Kosoe, E.A. (2014) 'Solid waste management in urban areas of Ghana: Issues and experiences from Wa', *Journal of Environment Pollution and Human Health*, 2(5), pp. 110–117.

Arias, O., Pulgar, J.A. and Soto, M. (2022) 'Application of organic wastes to soils and legislative intricacies in a circular economy context', *Clean Technologies and Environmental Policy*, 24(6), pp. 1871–1888.

Awino, F.B. and Apitz, S.E. (2024) 'Solid waste management in the context of the waste hierarchy and circular economy frameworks: An international critical review', *Integrated Environmental Assessment and Management*, 20(1), pp. 9–35.

Awuchi, C.G., Amagwula, I.O. and Igwe, V.S. (2020) 'Industrial and community waste management: Global perspective', *American Journal of Physical Sciences*, 1(1), pp. 1–16.

Boateng, S., Boakye-Ansah, D., Baah, A., Aboagye, B. and Kyreece, P.A.G. (2023) 'Solid waste management practices and challenges in rural and urban senior high schools in Ashanti Region, Ghana', *Journal of Environmental and Public Health*, 2023, p. 9694467.

Cheng, X. et al. (2024) 'Optimizing rural waste management: Leveraging high-resolution remote sensing and GIS for efficient collection and routing', *International Journal of Applied Earth Observation and Geoinformation*, 135, p. 104219.

Ddiba, D., Andersson, K., Rosemarin, A., Schulte-Herbrüggen, H. and Dickin, S. (2022) 'The circular economy potential of urban organic waste streams in low- and middle-income countries', *Environment, Development and Sustainability*, 24(1), pp. 1116–1144.

Dhokhikah, Y. and Trihadiningrum, Y. (2012) 'Solid waste management in Asian developing countries: Challenges and opportunities', *Journal of Applied Environmental and Biological Sciences*, 2(7), pp. 329–335.

El-Fadel, M., Zeinati, M., El-Jisr, K. and Jamali, D. (2001) 'Industrial-waste management in developing countries: The case of Lebanon', *Journal of Environmental Management*, 61(4), pp. 281–300.

Elia, V., Gnoni, M.G. and Tornese, F. (2020) 'Evaluating the adoption of circular economy practices in industrial supply chains: An empirical analysis', *Journal of Cleaner Production*, 273, p. 122966.

Ellacuriaga, M., García-Cascallana, J. and Gómez, X. (2021) 'Biogas production from organic wastes: Integrating concepts of circular economy', *Fuels*, 2(2), pp. 144–167.

Ezeudu, O.B. and Ezeudu, T.S. (2019) 'Implementation of circular economy principles in industrial solid waste management: Case studies from a developing economy (Nigeria)', *Recycling*, 4(4), p. 42.

Geng, Y., Tsuyoshi, F. and Chen, X. (2010) 'Evaluation of innovative municipal solid waste management through urban symbiosis: A case study of Kawasaki', *Journal of Cleaner Production*, 18(10–11), pp. 993–1000.

Guerrero, L.A., Maas, G. and Hogland, W. (2013) 'Solid waste management challenges for cities in developing countries', *Waste Management*, 33(1), pp. 220–232.

Haque, F., Fan, C. and Lee, Y.Y. (2023) 'From waste to value: Addressing the relevance of waste recovery to agricultural sector in line with circular economy', *Journal of Cleaner Production*, 415, p. 137873.

Henry, R.K., Yongsheng, Z. and Jun, D. (2006) 'Municipal solid waste management challenges in developing countries: Kenyan case study', *Waste Management*, 26(1), pp. 92–100.

Iqbal, A., Yasar, A., Tabinda, A.B., Haider, R., Sultan, I.A., Kedwii, A.A., Chaudhary, M.M., Sheikh, M.M. and Nizami, A.S. (2023) 'Waste as resource for Pakistan: An innovative business model of regenerative circular economy to integrate the municipal solid waste management sector', *Sustainability*, 15(7), p. 6281.

Iqbal, N., Agrawal, A., Dubey, S. and Kumar, J. (2020) 'Role of decomposers in agricultural waste management', in Basso, T.P., Basso, T.O. and Basso, L.C. (eds.), *Biotechnological Applications of Biomass*. Rijeka: IntechOpen. https://doi.org/10.5772/intechopen.94328

Izuchukwu Precious, O., Zino Izu, O., Theresa Ojevwe, A. and Frank Chudi, A. (2025) 'Smart cities and circular economy: Advancing waste management through urban innovation in Nigeria', *Journal of Economy, Tourism and Service*, 4(2), pp. 85–115.

Kaszycki, P., Głodniok, M. and Petryszak, P. (2021) 'Towards a bio-based circular economy in organic waste management and wastewater treatment: The Polish perspective', *New Biotechnology*, 61, pp. 80–89.

Kotyal, K. (2023) 'Circular agriculture: Sustainable farming practices for zero waste', *Environmental Reports*, 5(1), pp. 8–12. https://doi.org/10.51470/ER.2023.5.1.08

Kurniawan, T.A., Avtar, R., Singh, D., Xue, W., Othman, M.H.D., Hwang, G.H., Iswanto, I., Albadarin, A.B. and Kern, A.O. (2021) 'Reforming municipal solid waste management in Sukunan (Yogyakarta, Indonesia): Applying a zero-waste approach based on a circular economy paradigm', *Journal of Cleaner Production*, 284, p. 124775.

Leknoi, U., Painmanakul, P., Chawaloesphonsiya, N., Wimolsakcharoen, W., Samritthinanta, C. and Yiengthaisong, A. (2024) 'Building sustainable community: Insight from a successful waste management initiative', *Resources, Conservation and Recycling Advances*, 24, p. 200238.

Lindgren, G.F. (2018) *Managing Industrial Hazardous Waste—A Practical Handbook*. Boca Raton, FL: CRC Press.

Ling, Z.S., Jaafar, M.H. and Ismail, N. (2025) 'The knowledge, attitudes and practices in circular economy: A review of industrial waste management in China', *Cleaner Waste Systems*, 11, p. 100276. https://doi.org/10.1016/j.clwas.2025.100276

Maitre-Ekern, E. (2021) 'Re-thinking producer responsibility for a sustainable circular economy: From extended producer responsibility to pre-market producer responsibility', *Journal of Cleaner Production*, 286, p. 125454.

Maji, S. et al. (2020) 'Agricultural waste: Its impact on environment and management approaches', in Bharagava, R.N. (ed.), *Emerging Eco-Friendly Green Technologies for Wastewater Treatment: Microorganisms for Sustainability*. Singapore: Springer, pp. 329–351. https://doi.org/10.1007/978-981-15-1390-9_15

Masullo, A. (2017) 'Organic wastes management in a circular economy approach: Rebuilding the link between urban and rural areas', *Ecological Engineering*, 101, pp. 84–90.

Meng, M., Wen, Z., Luo, W. and Wang, S. (2021) 'Approaches and policies to promote zero-waste city construction: China's practices and lessons', *Sustainability*, 13(24), p. 13537.

Mihai, F.C. (2017) 'Introductory chapter: Rural waste management issues', in Mihai, F.-C. (ed.), *Solid Waste Management in Rural Areas*. Rijeka: IntechOpen, pp. 1–10. https://doi.org/10.5772/intechopen.70268

Mihai, F.C. et al. (2021) 'Plastic pollution, waste management issues, and circular economy opportunities in rural communities', *Sustainability*, 14(1), p. 20.

Mihai, F.C. (2023) 'Circular economy and sustainable rural development.' *Sustainability* 15(3), p. 2139.

Moharana, P.C. (2014) *Rural Solid Waste Management: Issues and Action*. Nagpur: National Bureau of Soil Survey and Land Use Planning.

Naqvi, S.R., Beig, B. and Naqvi, M. (2021) 'Circular economy approach to address industrial solid waste management', in Baskar, C., Ramakrishna, S., Baskar, S., Sharma, R., Chinnappan, A. and Sehrawat, R. (eds.), *Handbook of Solid Waste Management: Sustainability through Circular Economy*. Singapore: Springer, pp. 421–440. https://doi.org/10.1007/978-981-16-4230-2_16

Navia, R. and Bezama, A. (2008) 'Hazardous waste management in Chilean main industry: An overview', *Journal of Hazardous Materials*, 158(1), pp. 177–184.

Nemerow, N.L. (2010) *Industrial Waste Treatment: Contemporary Practice and Vision for the Future*. Amsterdam: Elsevier.

Nemerow, N.L. and Agardy, F.J. (1998) *Strategies of Industrial and Hazardous Waste Management*. New York: John Wiley & Sons.

Nwamekwe, C.O. and Chikwendu, O.C. (2025) 'Circular economy strategies in industrial engineering: From theory to practice', *International Journal of Multidisciplinary Research and Growth Evaluation*, 6(01), pp. 1773–1782.

Otundo Richard, M. (2024) 'Innovative approaches to solid waste management in Kenya's urban areas: A case study of Mombasa County in advancing the Sustainable Development Goals', *SSRN Electronic Journal*. Amsterdam: Elsevier (SSRN). https://doi.org/10.2139/ssrn.4937595

Oyedele, O. (2016) 'Challenges of urban solid waste management in developing countries', *Journal of Solid Waste Technology & Management*, 42(1), pp. 632–643.

Paes, M.X., de Oliveira, J.A.P., Mancini, S.D. and Rieradevall, J. (2024) 'Waste management intervention to boost circular economy and mitigate climate change in cities of developing countries: The case of Brazil', *Habitat International*, 143, p. 102990.

Rimantho, D., Suyitno, B.M., Pratomo, V.A., Haryanto, G., Prasidha, I.N.T. and Puspita, N. (2023) 'Circular economy: Barriers and strategy to reduce and manage solid waste in the rural area at Jepara district, Indonesia', *International Journal of Sustainable Development and Planning*, 18(4), pp. 1045–1055. https://doi.org/10.18280/ijsdp.180407.

Selvan, T. et al. (2023) 'Circular economy in agriculture: Unleashing the potential of integrated organic farming for food security and sustainable development', *Frontiers in Sustainable Food Systems*, 7, p. 1170380.

Schoeman, Y., Oberholster, P. and Somerset, V. (2021) 'A decision-support framework for industrial waste management in the iron and steel industry: A case study in Southern Africa', *Case Studies in Chemical and Environmental Engineering*, 3, p. 100097.

Sharma, V., Indora, A., Srivastava, A. and Singh, Y. (2025) 'Innovative approaches to solid waste management in rural communities: Challenges and solutions', in Albattat, A., Singh, A., Tyagi, P.K. and Haghi, A.K. (eds.) *Solid Waste Management for Rural Regions*. Hershey, PA: IGI Global Scientific Publishing, pp. 1–20. https://doi.org/10.4018/979-8-3693-8527-2.ch001

Singh, S. et al. (eds.) (2022) *Agriculture Waste Management and Bioresource: The Circular Economy Perspective*. Hoboken, NJ: John Wiley & Sons.

Srivastava, V., Ismail, S.A., Singh, P. and Singh, R.P. (2015) 'Urban solid waste management in the developing world with emphasis on India: Challenges and opportunities', *Reviews in Environmental Science and Bio/Technology*, 14(2), pp. 317–337.

Stentiford, E.I. et al. (2004) 'The importance of agriculture in global waste management', in Bernal, M.P., Moral, R., Clemente, R. and Paredes, C. (eds.), *Sustainable Organic Waste Management for Environmental Protection and Food Safety: Proceedings of the 11th International Conference of the FAO ESCORENA Network on Recycling of Agricultural, Municipal and Industrial Residues in Agriculture (RAMIRAN 2004)*. Murcia, Spain, 6–9 October 2004. Murcia: Food and Agriculture Organization of the United Nations (FAO) and Consejo Superior de Investigaciones Científicas (CSIC).

Tang, J. et al. (2020) 'A spatio-temporal perspective of China's industrial circular economy development', *Science of the Total Environment*, 706, p. 135754.

Tawo, O.E. (2025) 'Sustainable waste management strategies for circular economy in industrial operations and environmental protection', *International Journal of Research*, 6(1), pp. 3636–3652.

Tskhovrebov, E., Velichko, E. and Niyazgulov, U. (2019) 'Planning measures for environmentally safe handling with extremely and highly hazardous wastes', *Materials Science Forum*, 945, pp. 988–994. Trans Tech Publications.

Veckalne, R. and Tambovceva, T. (2021) 'Innovations in circular economy for sustainable urban development', *Marketing and Management of Innovations*, 4, pp. 196–209. https://doi.org/10.21272/mmi.2021.4-15.

Vinti, G. and Vaccari, M. (2022) 'Solid waste management in rural communities of developing countries: An overview of challenges and opportunities', *Clean Technologies*, 4(4), pp. 1138–1151.
Viva, L., Ciulli, F., Kolk, A. and Rothenberg, G. (2020) Designing circular waste management strategies: The case of organic waste in Amsterdam', *Advanced Sustainable Systems*, 4(9), p. 2000023.
Wang, A. et al. (2017) 'Rural solid waste management in China: Status, problems and challenges', *Sustainability*, 9(4), p. 506.
Xue, L. et al. (2016) 'Agricultural waste', *Water Environment Research*, 88(10), pp. 1334–1369.
Yadav, H., Soni, U. and Kumar, G. (2023) 'Analysing challenges to smart waste management for a sustainable circular economy in developing countries: A fuzzy DEMATEL study', *Smart and Sustainable Built Environment*, 12(2), pp. 361–384.
Zohoori, M. and Ghani, A. (2017) 'Municipal solid waste management challenges and problems for cities in low-income and developing countries', *International Journal of Science and Engineering Applications*, 6(2), pp. 39–48.

5 Case Studies and Best Practices

5.1 SUCCESSFUL IMPLEMENTATIONS

This chapter offers and analyzes empirical findings from a case study on CE implementation in a developing economy scenario. It analyzes the implementation of CE concepts in the context of resource restrictions, institutional constraints, and significant informality. Special emphasis is placed on operational processes, stakeholder engagement, policy coherence, and local innovation. The analysis correlates observed practices with the study objectives and conceptual framework, emphasizing context-specific facilitators and obstacles. This chapter offers substantive insights into the challenges of executing CE programs in underdeveloped nations, guiding practical and policy-focused recommendations in later chapters.

5.1.1 Implementation of CE in a Developing Economy's Mining Industry Using Institutional Theory: The Case of Namibia

The case study about the application of the CE within Namibia's mining sector illustrates how institutional theory can proficiently elucidate both the processes and results of sustainability transitions in developing economies. In Namibia, the adoption of CE did not arise solely from market pressures, but rather through the interplay of regulative, normative, and cognitive institutions. Regulatory pressures were pivotal: governmental policies for environmental protection, mine closure, waste management, and resource efficiency necessitated mining companies to reevaluate linear extraction methods. Mining businesses implemented procedures including waste rock reprocessing, water recycling, tailings reuse, and enhanced energy efficiency, especially in uranium and diamond mining operations. These activities were bolstered by normative demands from worldwide mining norms, foreign investors, and multinational parent firms, which increasingly demand adherence to ESG and CE concepts. Cognitive institutions, collective beliefs, and professional standards contributed significantly, as engineers, managers, and policymakers progressively reconceptualized waste as a resource and circularity to sustain competitiveness rather than merely a regulatory encumbrance. The CE transition succeeded in this instance because it was institutionally integrated rather than regarded as an isolated technological measure. Companies integrated circular processes with current legal frameworks, corporate sustainability objectives, and worldwide reporting standards, thereby diminishing resistance and augmenting legitimacy. The collaboration among government agencies, mining companies, local suppliers, and research institutions facilitated experimentation and learning, while export-oriented mining operations acquainted firms with international best practices and technologies. Nonetheless, the

DOI: 10.1201/9781003591764-5

story underscores limitations characteristic of emerging countries, such as inadequate local recycling infrastructure, talent deficiencies, and reliance on foreign technology suppliers. This encounter yields various lessons. Initially, the implementation of CE practices in resource-intensive sectors necessitates regulatory consistency and enforcement, rather than mere voluntary initiatives. Second, institutional alignment encompassing legislation, industry standards, and organizational attitudes is essential for maintaining circular processes. Third, external constraints from global value chains can serve as significant catalysts for circularity, while domestic capacity is still developing. The Namibian case exemplifies the importance of incorporating CE principles into mining licensing, environmental impact assessments, and closure plans, so ensuring circularity is addressed throughout the entire life cycle. Another applicable approach is industrial symbiosis, wherein mining waste streams are connected to construction materials, energy production, or secondary processing sectors. The case highlights that developing economies can facilitate CE transitions in extractive industries by strategically utilizing institutional frameworks, positioning Namibia as a pertinent model for other mineral-dependent nations pursuing sustainable and inclusive resource governance.

5.1.2 CE Business Models in Developing Economies: Lessons from India

The research conducted by Goyal et al. (2018) offers a thorough analysis of the emergence and operation of CE business models in developing economies, with India as the primary case study. The study illustrates that the application of the CE in developing environments is predominantly motivated by pragmatism, resource scarcity, and inclusive value creation, rather than only by environmental idealism. In contrast to industrialized economies, where circularity is frequently driven by legislation and technology, Indian circular business models have arisen organically based on the principles of reduce, recycle, and reuse (the 3Rs), addressing resource limitations, cost challenges, and unfulfilled social demands. The authors emphasize how companies, especially SMEs and social enterprises integrated circular practices into their fundamental business models by prolonging product life cycles, reclaiming resources, and devising cost-effective reuse mechanisms. Examples encompass remanufacturing of automobile components, recycling-oriented packaging solutions, repair and refurbishing operations, and community-driven waste recovery initiatives that use informal laborers. These approaches succeeded by integrating circular practices with cost-effectiveness, job creation, and local demand, so ensuring economic viability and social legitimacy. A primary factor for success was the robust connection between CE operations, and the capabilities of the informal sector, wherein waste pickers, repair technicians, and micro-entrepreneurs were instrumental in material collection and reuse. Instead of eliminating informality, effective CE models in India utilized it, progressively enhancing efficiency, safety, and income stability. The research indicates that corporate innovation, rather than regulation alone, was a principal catalyst, with companies exploring product–service systems, pay-per-use models, and decentralized recycling networks. Nonetheless, institutional deficiencies, including inadequate enforcement, restricted financial access, and disjointed policy frameworks, hindered expansion. This story yields multiple insights into developing economies. CE initiatives must be contextually aware, acknowledging local consumer behaviors, income strata, and institutional frameworks. Second, CE business models are more prone to success when they generate immediate economic value, including cost reductions or employment opportunities, in conjunction with environmental advantages. The informal sector ought to be regarded as an asset rather than an obstacle, with policies focused on progressive formalization, skill enhancement, and inclusiveness. The Indian experience presents replicable models, including repair-and-refurbish enterprises, materials recovery businesses associated with urban waste streams, reverse logistics collaborations with informal collectors, and cost-effective remanufacturing models for domestic markets. These models are especially pertinent for African and other developing countries experiencing swift urbanization, unemployment, and resource inefficiencies. The study substantiates the assertion that the adoption of a CE in developing nations is most efficacious when propelled by inclusive business models that integrate resource efficiency, social innovation, and market viability, rather than depending exclusively on top-down regulatory measures.

5.1.3 Implementation of CE Principles in Industrial SWM: Case Studies from a Developing Economy

The research conducted by Ezeudu and Ezeudu (2019) investigates the application of CE(CE) principles in the management of industrial solid waste in Nigeria, offering significant insights into the emergence of circularity within developing economic contexts, notwithstanding structural and institutional obstacles. The authors demonstrate through various industry case studies that in Nigeria, CE activities were predominantly motivated by cost reduction, material shortages, and regulatory compliance, rather than by explicit CE strategies. Manufacturing, food processing, and chemical industries implemented procedures such as waste segregation, on-site recycling, by-product recovery, and the reuse of industrial leftovers as secondary raw materials. In certain instances, industrial waste streams were diverted into other value chains; for instance, ashes, scrap metals, and organic wastes were repurposed for construction materials, energy production, or agricultural inputs. These approaches were effective primarily due to their provision of immediate operational advantages, including decreased disposal expenses, reduced raw material costs, and enhanced process efficiency. Nonetheless, execution was inconsistent and markedly distinctive to individual firms, indicative of poor national coordination, limited enforcement capabilities, and insufficient recycling infrastructure. The research emphasizes that the development of CE practices in Nigeria depended significantly on managerial commitment and internal innovation, rather than on systemic policy assistance. Companies that effectively adopted CE ideas allocated resources to staff training, process reengineering, and collaborations with local recyclers and informal garbage collectors. Simultaneously, other obstacles hindered wider adoption, such as inconsistent power supply, inadequate waste collection systems, restricted access to financing for cleaner technology, and insufficient awareness of CE principles among industrial stakeholders. This scenario yields various lessons. The creation of a CE in underdeveloped nations can commence inside current industrial frameworks by gradual enhancements, even without extensive CE legislation. Second, economic incentives are essential; companies are more inclined to embrace circular processes when cash savings and productivity enhancements are evident. Third, inadequate institutional frameworks result in CE outcomes frequently relying on individual enterprises instead of coordinated industrial ecosystems. The Nigerian example illustrates the possibility for industrial symbiosis at a local level, wherein the waste of one enterprise serves as input for another, facilitated by informal and semiformal recycling networks. A further transferable strategy is process-level circularity, encompassing internal reuse and recycling within manufacturing facilities, necessitating reduced capital investment, and allowing for incremental scalability. The study emphasizes the necessity of establishing supportive institutions, such as more defined waste rules, incentives for recycling investments, and platforms for information dissemination. The Nigerian case illustrates that although the implementation of a CE in industrial solid waste management within developing economies is hindered by systemic deficiencies, significant advancement is achievable when circular practices are integrated with firm-level economic incentives and bolstered by incremental institutional enhancement.

5.1.4 CE Model for Developing Countries: Evidence from Bangladesh

The research conducted by Ahmed et al. (2022) formulates and substantiates a CE(CE) model specifically designed for the contexts of developing nations, utilizing empirical data from Bangladesh. The research illustrates that the implementation of CE in developing nations is significantly influenced by structural limitations, socioeconomic priorities, and institutional capabilities, necessitating methods distinct from those utilized in developed situations. In Bangladesh, CE techniques have developed in manufacturing, textiles, construction, and waste management sectors, mostly motivated by resource scarcity, environmental degradation, and economic necessity rather than formal CE frameworks. The authors delineate essential CE mechanisms, including material reuse, recycling of industrial and municipal waste, energy recovery, eco-design, and cleaner manufacturing techniques, which were frequently implemented progressively and inconsistently across sectors. The model was effective as it incorporated economic performance, environmental sustainability, and social welfare, acknowledging that corporations and policymakers in developing nations emphasize job creation, cost efficiency, and poverty alleviation in conjunction with environmental objectives. Empirical investigation demonstrated that institutional quality, technological competence, green investment, and environmental rules positively impacted CE adoption, whereas inadequate governance, restricted financing, and insufficient awareness hindered advancement. A significant aspect of the Bangladesh instance is the prominent presence of SMEs and informal entities, especially in recycling, repair, and material recovery, which constitute the foundation of circular flows despite functioning outside established regulatory frameworks. The study indicates that the implementation of the CE was more effective when informal practices had indirect support from market demand, supply chain connections, and export-driven sectors like textiles, which saw pressure from overseas purchasers to adhere to sustainability criteria. The CE model put forth by the authors underscores the necessity of policy coordination, indicating that stand-alone environmental measures are inadequate without integration with industrial, energy, and social policies. This scenario yields several lessons. CE transitions in developing nations must prioritize development, clearly connecting circularity to employment, competitiveness, and inclusive growth. Second, enhancing institutional capacity, especially in regulatory enforcement, access to green financing, and knowledge transfer, is essential for expanding CE practices beyond pilot projects. Third, participation in the informal sector should be purposefully incorporated rather than overlooked, as it currently fulfills critical circular tasks. The Bangladesh experience provides transferable models, including sector-specific CE pathways (e.g., textiles-oriented recycling and water reuse), export-driven circular compliance frameworks that promote adoption through global value chain demands, and hybrid formal–informal systems that integrate grassroots recycling with formal industrial processing. The study additionally presents a systems-oriented CE paradigm that can assist policymakers in other developing nations in prioritizing interventions commencing with low-cost reuse and recycling, subsequently advancing to cleaner production and innovation as capacity expands. The Bangladesh story underscores that the adoption of a CE in developing nations is most efficacious when anchored in local economic conditions, bolstered by cohesive institutions, and conceived as a gradual, inclusive transformation rather than an abrupt technological advancement.

5.2 CHALLENGES AND FAILURES IN IMPLEMENTATION

5.2.1 Barriers and Challenges to Waste Management Hindering the CE in Sub-Saharan Africa

According to Debrah et al. (2022), various prevalent obstacles persistently hinder the adoption of the CE in sub-Saharan Africa, especially in waste management systems. A primary obstacle is inadequate institutional capacity, shown in disjointed governance frameworks, redundant mandates, and insufficient implementation of current waste legislation. Numerous towns lack the requisite technical expertise, financial resources, and infrastructure necessary for effective waste segregation, recycling, and recovery. Insufficient finance channels and restricted access to green investment impede both public agencies and private enterprises from embracing circular technologies. The study emphasizes significant informality in waste management, wherein informal participants are essential for material recovery yet are marginalized from official planning, financing, and regulatory structures. Insufficient public knowledge, adverse attitudes of recycled items, and inadequate data systems regarding waste flows further constrain the efficacy and scalability of circular programs.

These obstacles have led to several unsuccessful or suboptimal initiatives around the region. The publication outlines instances in which donor-funded recycling and waste-to-energy initiatives failed during trial periods, attributed to inadequate maintenance, absence of local ownership, and misalignment with local socioeconomic contexts. In several cities, centralized recycling facilities have failed due to inadequate source segregation of waste, leading to contamination and elevated processing expenses. Other projects faltered because of their dependence on foreign technologies that were costly to operate and ill-suited for local waste compositions. Moreover, efforts to formalize waste management systems without incorporating informal waste pickers frequently resulted in opposition, loss of livelihoods, and subsequent project abandonment. These failures illustrate that CE initiatives cannot thrive when they are technologically driven yet institutionally fragile or socially disengaged.

Debrah et al. (2022) contend that these failures present significant learning opportunities for enhancing future CE initiatives. A fundamental lesson is the necessity for context-sensitive design, wherein CE activities are tailored to local waste attributes, income strata, and government conditions. Enhancing institutions via improved policy coordination, reliable financial channels, and capacity development at the local level is crucial for sustained success. The study emphasizes the significance of inclusive strategies that acknowledge and incorporate informal actors as collaborators instead of impediments. Incremental transitions, commencing with cost-effective measures like source separation, community-oriented recycling, and repair initiatives, are more feasible than extensive, capital-heavy programs. Ultimately, deriving insights from previous failures underscores the necessity to transition from short-term, donor-dependent initiatives to locally integrated, economically sustainable circular models. By integrating learning, adaptability, and stakeholder engagement into policy and practice, sub-Saharan African nations can convert previous challenges into a basis for more robust and efficient CE implementation.

5.2.2 Challenges of the South African Economy to Transition to a CE: A Case of Remanufacturing

The research conducted by Ohiomah and Sukdeo (2022) investigates the obstacles encountered by South Africa in its shift towards a circular economy, focusing on remanufacturing, a fundamental circular strategy with considerable promise in developing and emerging economies. The authors demonstrate that, although South Africa possesses a reasonably robust industrial foundation compared to several African nations, the implementation of remanufacturing is constrained by structural, institutional, and market-related obstacles. Principal obstacles are erratic policy support; insufficient coordination among industrial, trade, and environmental policies; and the lack of definitive criteria and incentives for remanufactured products. The absence of legal frameworks diminishes investor trust and constrains the incorporation of remanufacturing within established manufacturing and supply chains.

The document additionally emphasizes market and demand-side limitations that hinder remanufacturing initiatives. Consumer perceptions of remanufactured products as inferior, unreliable, or low quality diminish market acceptance, especially in industries such as automobile components and industrial machinery. This view is exacerbated by insufficient awareness campaigns and the absence of labeling or certification systems that could indicate quality and safety. Moreover, original equipment manufacturers (OEMs) frequently oppose remanufacturing due to apprehensions regarding intellectual property, the potential to undermine new product sales, and inadequate reverse logistics systems for product retrieval. These limitations restrict access to cores (used items) necessary for remanufacturing, thereby elevating operational costs and uncertainty.

The study cites skills shortages, technology limits, and infrastructural deficiencies as significant hurdles from an operating standpoint. Remanufacturing necessitates specialized technical competence, diagnostic instruments, and quality assurance procedures, which are either limited or expensive in the South African setting. Small and medium firms, which could normally propel remanufacturing innovation, encounter restricted access to financing and elevated compliance expenses. Consequently, numerous remanufacturing endeavors remain informal or limited in scale, hindering their capacity to expand and assimilate into established industrial ecosystems. The authors additionally observe that elevated energy expenses and inconsistent infrastructure further undermine the economic feasibility of remanufacturing activities.

The South African case provides numerous insights for CE transitions in poor nations. First, remanufacturing cannot prosper without cohesive and supportive legislative frameworks that acknowledge remanufactured items within industrial and trade regulations. Second, fostering demand via standards, certification, and public procurement is crucial for enhancing consumer confidence and market adoption. Third, successful circular transitions necessitate robust reverse logistics systems and collaboration throughout value chains, encompassing OEMs, suppliers, and independent remanufacturers. The study highlights the possibility of repeatable methods such as public–private partnerships, skills development programs, and incentives for SMEs to formalize and expand remanufacturing activities. The research underscores that although remanufacturing presents significant potential as a CE strategy in underdeveloped nations, its efficacy relies on intentional institutional support, market growth, and capacity building, rather than solely on technical feasibility.

5.2.3 Barriers Toward CE Transition: Exploring Different Stakeholders' Perspectives

The study by Souza Piao et al. (2024) provides a comprehensive and multidimensional analysis of barriers to CE transition, emphasizing that circularity challenges vary significantly across stakeholder groups, including policymakers, businesses, supply chain actors, consumers, and civil society. Unlike studies that focus on single-sector or firm-level constraints, this research demonstrates that CE implementation failures often stem from misaligned stakeholder expectations, incentives, and capabilities, rather than from technical limitations alone. From a policy perspective, key barriers include fragmented regulations, lack of long-term policy coherence, and weak coordination across ministries and governance levels. Even where CE strategies exist, unclear responsibilities and inconsistent enforcement undermine confidence among private actors, slowing investment and experimentation. From the business and industry perspective, the study identifies economic and operational barriers as dominant. Firms face high upfront costs of circular redesign, limited access to finance, uncertain return on investment, and underdeveloped markets for secondary materials. Many companies, especially small and medium enterprises (SMEs), struggle with insufficient technical knowledge, lack of skilled Labour, and poor access to enabling infrastructure such as reverse logistics systems and recycling facilities. The authors also highlight supply chain-related barriers, where linear value chains, power asymmetries, and limited collaboration restrict information sharing and joint circular initiatives. In developing economy contexts, these challenges are often exacerbated by informality, weak supplier capabilities, and dependence on imported technologies. The consumer and societal perspective reveal another critical layer of resistance. Souza Piao et al. (2024) show that low awareness of CE concepts, cultural preferences for new products, and distrust of reused or remanufactured goods significantly reduce demand for circular products and services. These perceptions weaken market pull and discourage firms from scaling circular innovations. Civil society actors further point to social equity concerns, warning that poorly designed CE transitions risk, excluding informal workers, or shifting environmental burdens rather than reducing them. This reinforces the idea that CE is not only a technical or economic transition but also a social and behavioral transformation. A key contribution of the study lies in its analysis of why CE initiatives stall or fail. The authors argue that failures often occur when CE policies are designed top-down without stakeholder engagement, when business models are promoted without consumer acceptance, or when circular technologies are introduced without supportive institutions. The lack of shared understanding and coordinated action leads to fragmented pilots, limited scalability, and eventual loss of momentum. Importantly, the paper reframes these failures as opportunities for learning. It emphasizes the need for stakeholder-inclusive governance, cross-sector dialogue, and adaptive policy frameworks that evolve through experimentation and feedback. The lessons from this study are highly relevant for developing economies. First, CE transitions require systemic alignment policies, markets, technologies, and social norms must evolve together. Second, stakeholder engagement should move beyond consultation to co-creation, ensuring that circular solutions are context-appropriate and socially legitimate. Third, gradual and hybrid transition pathways combining formal and informal systems, and public and private actors are more realistic than rapid, uniform transformations. Overall, Souza Piao et al. (2024) reinforce the argument that successful CE implementation depends less on technological readiness and more on governance quality, stakeholder coordination, and shared value creation, offering critical insights for designing more resilient and inclusive CE strategies in both developing and emerging economies.

5.2.4 Implementing CE Practices in Indian SMEs: Analysis of Challenges with Case Study

Kumar et al. (2025) conduct a comprehensive examination of the obstacles to implementing CE practices in Indian SMEs, substantiated by empirical case study evidence. The report emphasizes that although SMEs are crucial for promoting circularity in emerging economies because of their adaptability and closeness to local markets, they encounter significant structural and operational challenges. Identified key constraints encompass restricted financial resources, inadequate access to cost-effective green technology, and a deficiency in technical expertise concerning circular design, waste valorization, and reverse logistics. In contrast to large enterprises, SMEs frequently function with narrow profit margins, rendering long-term investments in circular processes precarious in the absence of evident short-term financial rewards.

The case study evidence indicates that the CE techniques implemented by Indian SMEs such as material reuse, process optimization, waste reduction, and basic recycling were predominantly incremental and focused on efficiency, rather than being transformative. These approaches were effective when they coincided with cost reduction, regulatory adherence, or consumer demands, especially within manufacturing clusters associated with larger corporations or export markets. The lack of organized CE frameworks, inadequate data on material flows, and insufficient coordination throughout supply chains considerably hindered scalability. The research identifies institutional obstacles, such as disjointed governmental support, insufficient incentives designed for SMEs, and inadequate incorporation of CE aims into SME development initiatives. Consequently, numerous circular initiatives persisted as discrete pilot projects instead of being integrated into overarching business objectives.

An essential finding from the article is the significance of managerial awareness and leadership. Small and medium-sized enterprises with proprietors or managers knowledgeable about CE ideas were more inclined to engage in circular practices, especially in challenging circumstances. In contrast, insufficient understanding and skill deficiencies resulted in resistance, misperception of CE as an additional expense, or total inaction. The study also highlights problems with market adoption, as buyers exhibited hesitance to purchase recycled or remanufactured items, hence diminishing incentives for SMEs to invest in circular innovation. Infrastructure-related impediments such as insufficient garbage collection systems and inconsistent electricity supply further hindered implementation.

The study yields multiple lessons and transferable insights. Initially, the adoption of CE in SMEs necessitates specific governmental instruments, such as fiscal incentives, subsidized technologies, and specialized CE capacity-building initiatives. Second, cluster-based strategies and collaborations with larger corporations, academic institutions, and industry groups can assist SMEs in surmounting scale and knowledge constraints. Third, a phased implementation initiating with cost-effective efficiency and reuse strategies prior to progressing to more intricate circular models is more feasible in the context of emerging economies. Kumar et al. (2025) illustrate that although Indian SMEs encounter substantial obstacles to CE adoption, these impediments can be alleviated through supportive institutions, supply chain integration, and pragmatic, context-specific implementation strategies, providing insightful lessons for SME-led circular transitions in other developing economies.

REFERENCES

Ahmed, Z., Mahmud, S. and Acet, H. (2022) 'Circular economy model for developing countries: Evidence from Bangladesh', *Heliyon*, 8(5), e09580. https://doi.org/10.1016/j.heliyon.2022.e09580

Debrah, J.K., Teye, G.K. and Dinis, M.A.P. (2022) 'Barriers and challenges to waste management hindering the circular economy in Sub-Saharan Africa', *Urban Science*, 6(3), p. 57. https://doi.org/10.3390/urbansci6030057

Ezeudu, O.B. and Ezeudu, T.S. (2019) 'Implementation of circular economy principles in industrial solid waste management: Case studies from a developing economy (Nigeria)', *Recycling*, 4(4), p. 42. https://doi.org/10.3390/recycling4040042

Goyal, S., Esposito, M. and Kapoor, A. (2018) 'Circular economy business models in developing economies: Lessons from India on reduce, recycle and reuse paradigms', *Thunderbird International Business Review*, 60(5), pp. 729–740. https://doi.org/10.1002/tie.21917

Kahupi, I., Yakovleva, N., Okorie, O. and Hull, C.E. (2024) 'Implementation of circular economy in a developing economy's mining industry using institutional theory: The case of Namibia', *Journal of Environmental Management*, 368, Article 122145. https://doi.org/10.1016/j.jenvman.2024.122145

Kumar, S., Kumar, R., Phanden, R.K. and Rathi, R. (2025) 'Implementing circular economy practices in Indian SMEs: Analysis of challenges with a case study', *International Journal of Advanced Operations Management*, 16(1), pp. 35–59.

Ohiomah, I. and Sukdeo, N. (2022) 'Challenges of the South African economy in transitioning to a circular economy: The case of remanufacturing', *Journal of Remanufacturing*, 12(2), pp. 213–225. https://doi.org/10.1007/s13243-021-00108-z

Souza Piao, R., Vincenzi, T.B.D., Vazquez-Brust, D.A., Yakovleva, N., Bonsu, S. and Carvalho, M.M.D. (2024) 'Barriers toward circular economy transition: Exploring different stakeholders' perspectives', *Corporate Social Responsibility and Environmental Management*, 31(1), pp. 153–168. https://doi.org/10.1002/csr.2592

6 The Future of Circular Economy in Solid Waste Management

6.1 OVERVIEW

The future of CE in SWM is increasingly influenced by the intersection of environmental mandates, digital advancements, and inclusive development strategies, especially in developing nations. Recent studies indicate that CE serves not merely as an environmental strategy but as a comprehensive framework for attaining the UN Sustainable Development Goals, bolstering resilience, and facilitating postcrisis recovery. Sharma et al. (2021) emphasize that post-COVID recovery offers a chance to reconfigure waste systems to focus on resource recovery, decentralized treatment, and the creation of green jobs, transitioning SWM from disposal-centric models to value retention systems. Mandpe et al. (2023) underscore that in developing economies, the future of CE depends on context-specific solutions that prioritize source waste segregation, cost-effective recycling technologies, composting, and the incorporation of informal actors that already fulfill critical circular roles. Simultaneously, technological transformation is reconfiguring the realm of possibilities. Evidence from China indicates that digitalization via smart bins, data-driven collection systems, traceability platforms, and digital marketplaces for secondary materials can markedly improve efficiency, transparency, and circular value creation when bolstered by robust institutions and investment (Kurniawan, 2022). The future of CE in SWM will hinge on the alignment of policy consistency, digital tools, behavioral change, and inclusive governance. Successful transitions will necessitate progressive implementation, stakeholder engagement, and the integration of CE ideas across urban planning, industrial policy, and community systems, rather than pursuing capital-intensive, uniform solutions. These studies collectively indicate that the future of CE in SWM is attainable, scalable, and revolutionary, contingent upon social inclusivity, digital facilitation, and institutional support.

6.2 TRENDS AND EMERGING OPPORTUNITIES

6.2.1 Future Trends in SWM and CE

6.2.1.1 Digitalization and Industry 4.0-Enabled CE Systems

A prominent future trend in SWM is the incorporation of digital technologies and Industry 4.0 capabilities to facilitate migrations to a CE. Research highlights the increasing significance of sensors, the IoT, AI, and data analytics in enhancing waste

DOI: 10.1201/9781003591764-6

collection efficiency, material tracking, and decision-making (Okorie et al., 2018; Cheah et al., 2022). Intelligent bins, real-time monitoring systems, and predictive analytics are anticipated to optimize collection routes, diminish operating expenses, and improve transparency throughout waste value chains. Digital platforms enhance the traceability of resources, hence promoting recycling, remanufacturing, and EPR initiatives. With the advancement of digital maturity, SWM systems are anticipated to transition from reactive disposal models to proactive, data-driven circular systems.

6.2.1.2 Green Technologies and Resource Recovery Innovations

Future CE-oriented SWM systems will increasingly depend on sustainable and environmentally friendly technologies that emphasize resource recovery rather than disposal. Tanveer et al. (2022) emphasize that developments in W-t-E, AD, bio-composting, and sophisticated recycling technologies are pivotal for attaining environmental sustainability. These technologies facilitate the retrieval of energy, nutrients, and secondary raw materials, thereby diminishing reliance on landfills and curtailing greenhouse gas emissions. Um (2025) also observes that modular and decentralized treatment methods are becoming increasingly significant, especially in fast urbanizing areas, as they diminish infrastructure expenses and improve resilience. The future direction of SWM is thus rooted in the integration of technological innovation with circular design concepts that optimize material and energy cycles.

6.2.1.3 CE Integration in Urban Waste Governance

Urbanization is transforming the future of SWM, rendering cities essential arenas for the implementation of CE. Sesay and Fang (2025) contend that forthcoming municipal waste systems must incorporate CE principles into urban planning, infrastructure development, and service delivery frameworks. This includes the design of urban environments for source waste segregation, the advancement of circular procurement practices, and the incorporation of informal waste stakeholders into formal systems. Innovations in governance via PPP, incentive-driven legislation, and participatory planning are anticipated to be pivotal in the expansion of circular solutions. Future SWM systems will increasingly implement citywide circular frameworks that integrate waste management with housing, transportation, energy, and industrial development, rather than relying on isolated pilot programs.

6.2.1.4 Behavioral Change, Skills, and Inclusive Circular Transitions

Technological advancement alone will not yield circular results without concomitant behavioral and institutional transformation. Numerous studies emphasize that future CE-oriented SWM systems must prioritize public awareness, skills enhancement, and stakeholder involvement (Tanveer et al., 2022; Um, 2025). Consumer engagement in waste segregation, acceptance of recycled goods, and implementation of reuse strategies will continue to be essential success determinants. In developing economies, inclusive models acknowledging the contributions of informal waste workers are increasingly regarded as vital for equitable and effective transformations. The future of CE in SWM relies on the integration of technology innovation with social inclusion and capacity development.

6.2.1.5 Integrated and Adaptive Circular Waste Systems

Future trends indicate the emergence of integrated, adaptable, and digitally facilitated circular waste systems. Literature indicates that effective transitions necessitate cohesive policies, cross-sector coordination, and long-term investment strategies that reconcile environmental objectives with economic and social considerations. Instead of linear end-of-pipe solutions, future SWM systems will operate as dynamic circular ecosystems, promoting sustainability, resilience, and equitable growth.

6.3 EMERGING OPPORTUNITIES FOR DEVELOPING ECONOMIES

6.3.1 Entrepreneurship and Innovation as Engines of CE Growth

Developing economies are increasingly finding opportunities inside the CE through entrepreneurship and innovation-driven growth. Panait et al. (2022) assert that CE principles foster new enterprises by converting waste, inefficiency, and resource scarcity into avenues for value creation. In emerging economies, characterized by unmet demands and institutional deficiencies, entrepreneurs are distinctly equipped to implement circular business models centered on reuse, repair, remanufacturing, recycling, and product–service systems. These models frequently arise not solely from environmental causes but as pragmatic answers to cost pressures, unemployment, and market instability. The CE thus provides developing nations with the chance to bypass traditional linear development models by integrating sustainability into business formation, especially within SMEs and informal sector innovators.

6.3.2 Digital Platforms and Technology-Enabled Circular Opportunities

Digitalization is rapidly improving CE opportunities in developing countries by lowering transaction costs, improving coordination, and enabling scalability. Nemilentseva et al. (2024) assert that digital platforms including online marketplaces, sharing platforms, traceability systems, and data-driven logistics tools enhance circular value chains by connecting waste producers, recyclers, manufacturers, and consumers. These platforms enable material recovery, reverse logistics, and peer-to-peer trading, even in settings with insufficient physical infrastructure. Digital CE solutions offer a relatively accessible entry point for poor nations, particularly in environments with substantial mobile technology utilization. The potential transcends technology; it includes platform-based entrepreneurship, where local innovators develop contextually pertinent digital solutions that integrate informal players and local marketplaces into circular ecosystems.

6.3.3 Sector-Specific Opportunities Across Developing Economies

An important finding from the research is that CE potential in developing nations is sector-specific rather than homogeneous. Godfrey et al. (2021) illustrate that in sectors including mining, agriculture, construction, manufacturing, and waste management, circular solutions can generate economic, environmental, and social values. In resource-intensive industries, opportunities arise in by-product valorization, IS, and asset life cycle extension. Circular approaches in agriculture and food systems, including composting, bioenergy, and nutrient recovery, augment productivity and mitigate environmental deterioration. Urban industries, especially construction and municipal waste, present significant opportunities for material recovery, secondary material reuse, and local job creation. These sectoral paths enable developing nations to synchronize CE implementation with current economic frameworks and comparative advantages.

6.3.4 Collaboration and Networks as Enablers of Circular Value

Collaboration serves as a vital facilitator of circular opportunities in developing environments. Mishra et al. (2021) demonstrate that the implementation of the CE is much enhanced by collaboration among enterprises, governments, communities, and intermediaries across value chains. In settings marked by scarce resources and disjointed institutions, collaborations facilitate the surmounting of scale limitations, knowledge deficiencies, and financial obstacles. Collaboration facilitates common infrastructure, collective learning, and synchronized innovation, especially for SMEs that lack the capacity to individually engage in circular technology. Developing economies can capitalize on the establishment of multi-stakeholder circular networks that connect official and informal participants, local businesses and global corporations, as well as public and private entities.

6.3.5 Systems Thinking and Integrated Circular Transitions

Emerging opportunities in developing economies are contingent upon embracing a systems perspective of the circular economy. Scheel and Aguiñaga (2025) contend that CE should be perceived not as a collection of discrete acts but as an integrated system encompassing production, consumption, governance, culture, and ecosystems. This system's perspective for developing countries uncovers potential to integrate CE principles with overarching development objectives, including industrialization, urban resilience, and social inclusion. Systems thinking facilitates the formulation of policies and business models that prevent burden-shifting and instead produce co-benefits across economic, environmental, and social spheres. This method is especially pertinent when developmental difficulties are intricately linked, necessitating coordinated and adaptable solutions.

6.3.6 Employment, Inclusion, and Social Value Creation

The CE presents a significant opportunity for employment development and social inclusion in emerging economies. Clube (2022) critically evaluates the legitimacy of employment development as a benefit of the CE and concludes that, although not guaranteed, job creation can be a substantial result when circular methods are labor-intensive and locally integrated. Activities including repair, refurbishing, recycling, and decentralized waste management typically produce a higher number of jobs per unit of material compared to linear systems. For emerging economies experiencing elevated unemployment and informality, the CE presents potential to formalize livelihoods, increase working conditions, and improve income stability especially for youth and marginalized populations. Nonetheless, actualizing this potential necessitates intentional policy decisions that emphasize inclusive circular paths rather than only capital-intensive solutions.

6.4 ROLE OF RESEARCH AND INNOVATION

Research and innovation are essential and transformative for promoting the CE, especially in developing and emerging nations pursuing sustainable development pathways. Research offers the foundational data required to comprehend material flows, environmental consequences, socioeconomic trade-offs, and institutional limitations within current linear systems. Research utilizes life cycle evaluation, material flow analysis, and systems modeling to pinpoint inefficiencies, leakage points, and potential for value retention within production and consumption systems. This knowledge is essential for formulating context-specific circular tactics instead of adopting models that may not correspond with local conditions. Innovation converts research findings into pragmatic solutions, encompassing novel technology, business models, governance frameworks, and societal practices that facilitate circularity. Technological innovation propels progress in recycling, remanufacturing, bio-based materials, waste-to-resource methodologies, and digital platforms that facilitate traceability and reverse logistics. Concurrently, non-technological innovation encompassing business model innovation, policy experimentation, and social innovation equally contributes to the integration of circular concepts within markets and communities. Research institutions and universities serve as centers for experimentation, pilot projects, and skill enhancement, promoting collaboration among industry, government, and society. In developing economies, research and innovation are essential for bypassing traditional development trajectories, allowing nations to implement low-carbon, resource-efficient systems without following environmentally detrimental paths. Innovation ecosystems that facilitate start-ups, SMEs, and informal entities can provide locally based circular solutions that create employment, boost livelihoods, and bolster resilience. Furthermore, research facilitates policy learning and adaptive governance, guiding legislation, standards, and incentives that promote circular transitions while mitigating unexpected social or environmental repercussions. As the CE rapidly converges with digitalization, Industry 4.0, and sustainability initiatives, ongoing research and innovation are vital for navigating complexity, scaling effective pilots, and guaranteeing inclusivity. The CE cannot advance beyond fragmented initiatives without continuous investment in research and innovation that integrates scientific knowledge, technological capacity, and societal need, establishing them as fundamental components of enduring circular change.

REFERENCES

Cheah, C.G., Chia, W.Y., Lai, S.F., Chew, K.W., Chia, S.R. and Show, P.L. (2022) 'Innovation designs of Industry 4.0 based solid waste management: Machinery and digital circular economy', *Environmental Research*, 213, p. 113619. https://doi.org/10.1016/j.envres.2022.113619

Clube, R. (2022) 'Is job creation a legitimate social benefit of the circular economy?' *Resources, Conservation and Recycling*, 181, p. 106220. https://doi.org/10.1016/j.resconrec.2022.106220

Godfrey, L.K., Nahman, A., Oelofse, S.H., Trotter, D., Khan, S., Nontso, Z., Magweregwede, F., Sereme, B.V., Okole, B.N., Gordon, G.E. and Brown, B. (2021) 'The CEas development opportunity: Exploring CEopportunities across South Africa's economic sectors', *South African Journal of Science*, 117(3/4), pp. 1–11. https://doi.org/10.17159/sajs.2021/8169

Kurniawan, I.G.A. (2022). 'Digitalization of business law: Urgency and orientation of the industrial revolution 4.0 and society 5.0' *Volksgeist: Jurnal Ilmu Hukum Dan Konstitusi*, pp. 253–265.

Mandpe, A., Paliya, S., Gedam, V.V., Patel, S., Tyagi, L. and Kumar, S. (2023) 'Circular economy approach for sustainable solid waste management: A developing economy perspective', *Waste Management & Research*, 41(3), pp. 499–511. https://doi.org/10.1177/0734242X221126718

Mishra, J.L., Chiwenga, K.D. and Ali, K. (2021) 'Collaboration as an enabler for circular economy: A case study of a developing country', *Management Decision*, 59(8), pp. 1784–1800. https://doi.org/10.1108/MD-10-2018-1111

Nemilentseva, M., Tariq, A., Tariq, W., Aghajani, D. and Torkkeli, M. (2024) 'A review of digital platform and circular economy: opportunities and challenges for developing countries', in *Human Perspectives of Industry 4.0 Organizations*. Cham: Springer, pp. 38–67. https://doi.org/10.1201/9781032616810-3

Okorie, O., Salonitis, K., Charnley, F., Moreno, M., Turner, C. and Tiwari, A. (2018) 'Digitisation and the circular economy: a review of current research and future trends', *Energies*, 11(11), p. 3009. https://doi.org/10.3390/en11113009

Panait, M., Hysa, E., Raimi, L., Kruja, A. and Rodriguez, A. (2022) 'Guest editorial: CEand entrepreneurship in emerging economies: Opportunities and challenges', *Journal of Entrepreneurship in Emerging Economies*, 14(5), pp. 673–677. https://doi.org/10.1108/JEEE-10-2022-0487

Scheel, C. and Aguiñaga, E. (2025) 'A systems view of circular economy', *Sustainability*, 17(3), p. 1268. https://doi.org/10.3390/su17031268

Sesay, R.E.V. and Fang, P. (2025) 'CEin municipal solid waste management: Innovations and challenges for urban sustainability', *Journal of Environmental Protection*, 16(2), pp. 35–65.

Sharma, H.B., Vanapalli, K.R., Samal, B., Cheela, V.S., Dubey, B.K. and Bhattacharya, J. (2021) 'Circular economy approach in solid waste management system to achieve UN Sustainable Development Goals: Solutions for post-COVID recovery', *Science of the Total Environment*, 800, Article 149605. https://doi.org/10.1016/j.scitotenv.2021.149605

Tanveer, M., Khan, S.A.R., Umar, M., Yu, Z., Sajid, M.J. and Haq, I.U. (2022) 'Waste management and green technology: Future trends in CEleading towards environmental sustainability', *Environmental Science and Pollution Research*, 29(53), pp. 80161–80178. https://doi.org/10.1007/s11356-022-23238-8

Um, N. (2025) 'Sustainable waste management in the context of the circular economy', *Sustainability*, 17(5), p. 1937. https://doi.org/10.3390/su17051937

7 Roadmap for Developing Economies

7.1 OVERVIEW

Implementing CE ideas in developing nations necessitates a gradual, system-focused, and context-aware approach that harmonizes economic development goals with environmental sustainability and social inclusion. Literature review evidence indicates that CE transitions are most effective when they integrate policy coherence, innovation, collaboration, and inclusive involvement, rather than depending exclusively on technology-driven solutions. The strategic action plan commences with institutional alignment, guaranteeing the incorporation of CE objectives into national development plans, industrial strategies, urban policies, and waste management frameworks. Explicit legislative signals, standards, and incentives mitigate uncertainty and promote private sector investment in circular solutions.

The second strategic pillar emphasizes the facilitation of circular business models and entrepreneurship. Emerging economies gain advantages by fostering small and medium-sized organizations, start-ups, and informal businesses engaged in reuse, repair, recycling, remanufacturing, and sharing-based models. This encompasses financial access, incubator assistance, skills enhancement, and market development via public procurement. Cluster-oriented and sector-specific strategies such as circular construction, agro-waste valorization, and urban recycling systems enable nations to leverage existing economic frameworks instead of adopting inappropriate models from abroad.

Third, digitalization and innovation must be effectively utilized to address infrastructure and coordination difficulties. Digital platforms, data-centric waste systems, and Industry 4.0 technologies enhance material traceability, reverse logistics, and resource efficiency. Digital solutions must be crafted for affordability and inclusivity, guaranteeing the engagement of SMEs and informal entities. Research and innovation systems are essential for testing, modifying, and expanding technology appropriate for local circumstances.

The fourth aspect is inclusive social integration, acknowledging that informal labor, communities, and consumers are pivotal to circular material flows. Formalizing while incorporating informal actors, fostering behavioral change, and enhancing working circumstances are crucial for socially sustainable transformations. Job creation, skills enhancement, and livelihood security must be clearly defined outputs of the CE, rather than merely presumed ancillary benefits.

The action plan underscores the importance of collaboration and systems governance. The implementation of a CE necessitates synchronized efforts among government, industry, education, finance, and society. Multi-stakeholder platforms, public–private collaborations, and adaptable policy frameworks facilitate learning,

DOI: 10.1201/9781003591764-7

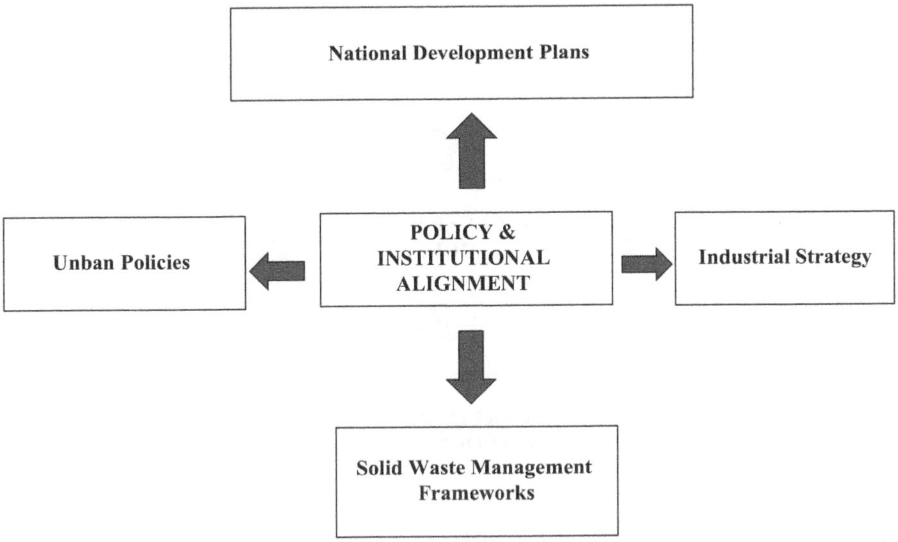

FIGURE 7.1 Policy and institutional alignment pillar

feedback, and enduring resilience. This strategy plan portrays the CE as a developmental pathway that enhances competitiveness, inclusivity, and sustainability in developing economies, rather than merely an environmental supplement.

7.1.1 Pillar 1: Policy and Institutional Alignment

This pillar constitutes the foundational framework of the action plan. It signifies the congruence of CE goals with national development agendas, industry strategies, urban policies, and SWM systems. Successful CE transitions in developing nations rely on cohesive rules, explicit institutional mandates, and uniform enforcement. Policies including EPR, green public procurement, waste segregation regulations, and incentives for recycling and remanufacturing establish the regulatory certainty essential for investment and innovation.

Figure 7.1 illustrates that this pillar is interconnected with the others, demonstrating how policy consistency facilitates circular business models, promotes digital innovation, and ensures social inclusion. In the absence of institutional cohesion, CE initiatives often remain disjointed, reliant on donors, or ephemeral.

7.1.2 Pillar 2: Circular Business Models and Entrepreneurship

This pillar signifies execution and value generation. It emphasizes SMEs, start-ups, and existing companies implementing circular business models, including reuse, repair, remanufacturing, recycling, and product–service systems. In developing economies, entrepreneurship serves as a vital catalyst for circularity, as enterprises frequently arise in reaction to scarcity, unemployment, and unfulfilled market demands.

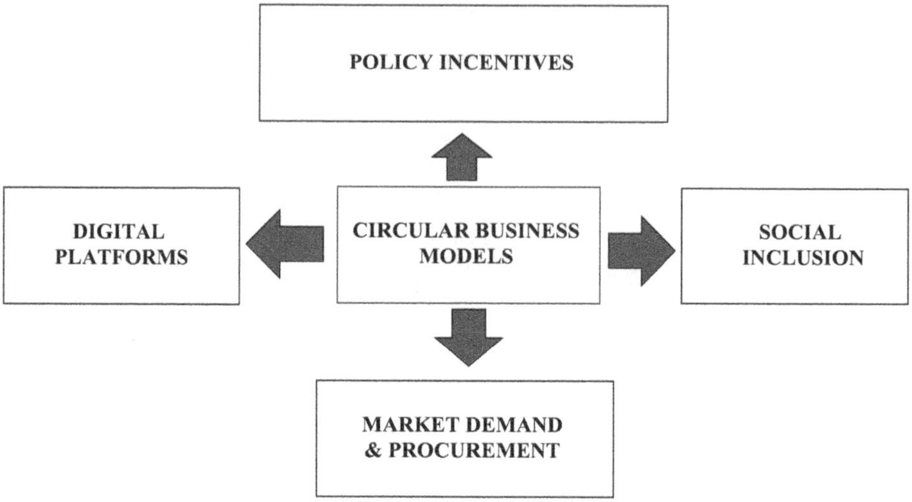

FIGURE 7.2 Circular business models and entrepreneurship pillar

Figure 7.2 illustrates the robust interaction of this pillar with policy (by incentives and procurement), digitalization (through platforms and data), and social inclusion (through employment and supplier networks). Circular business models convert CE ideas into economic activities, guaranteeing financial sustainability and scalability.

7.1.3 Pillar 3: Digitalization, Research, and Innovation

This pillar facilitates scalability, efficiency, and knowledge acquisition. It includes Industry 4.0 technologies, digital platforms, data systems, and research institutes that facilitate material traceability, reverse logistics, and evidence-based decision-making. Digitalization mitigates coordination flaws prevalent in emerging economies, while research and innovation tailor circular solutions to specific local circumstances.

Figure 7.3 illustrates bidirectional arrows representing ongoing feedback among innovation, policy refinement, and business acceptance. This pillar guarantees that CE transitions are flexible, technologically suitable, and based on local expertise rather than external solutions.

7.1.4 Pillar 4: Social Inclusion and Skills Development

This pillar guarantees the social validity and sustainability of circular transformations. It embodies communities, informal sector participants, customers, and the labor force. The integration of informal waste workers, enhancement of skills, modification of behaviors, and promotion of public awareness are crucial for maintaining circular material flows and averting social marginalization.

Figure 7.4 emphasizes robust connections between this pillar and circular business models (employment and livelihoods), in addition to policy (labor standards, education, and social protection). In the absence of social participation, the CE risks

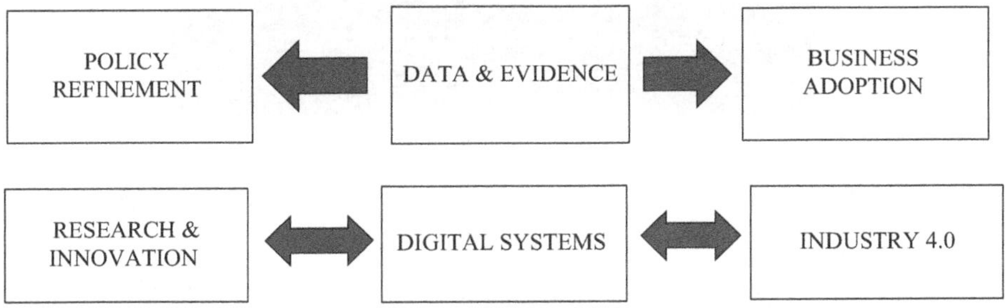

FIGURE 7.3 Digitalization, research and innovation pillar

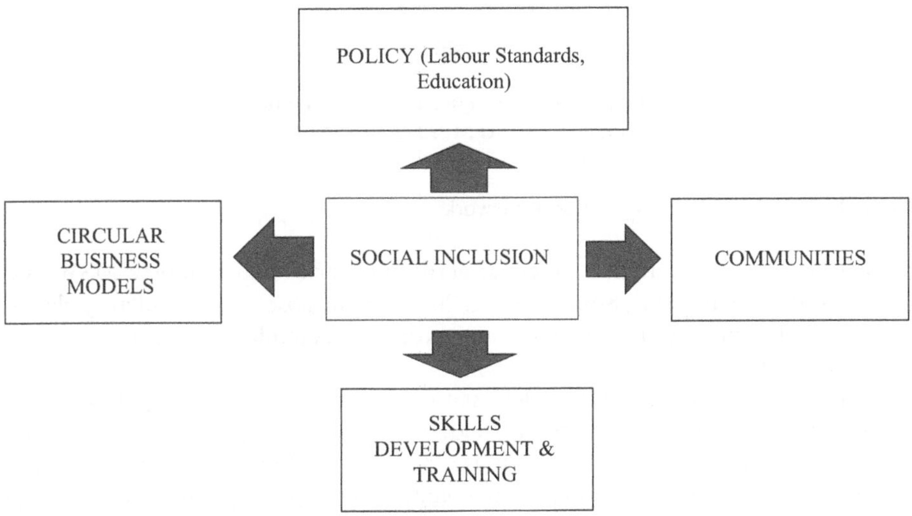

FIGURE 7.4 Social inclusion and skills development pillar

becoming capital-intensive and inequitable, thereby compromising its developmental potential.

7.2 INTEGRATED FOUR-PILLAR FRAMEWORK

The comprehensive four-pillar framework (Figure 7.5) offers a systems-based strategy for promoting CE transitions in developing nations, identifying SWM as a pivotal entry point for sustainable development. CE transitions should be perceived as interrelated socio-technical transformations rather than as stand-alone environmental reforms, according to systems thinking approaches (Scheel & Aguiñaga, 2025). The framework merges governance, enterprise, innovation, and social development into a cohesive system.

Pillar 1: Policy and institutional alignment establish the foundational framework. Coherent laws, such as EPR, sustainable public procurement, and waste requirements, mitigate uncertainty and encourage investment. Coordinated policy

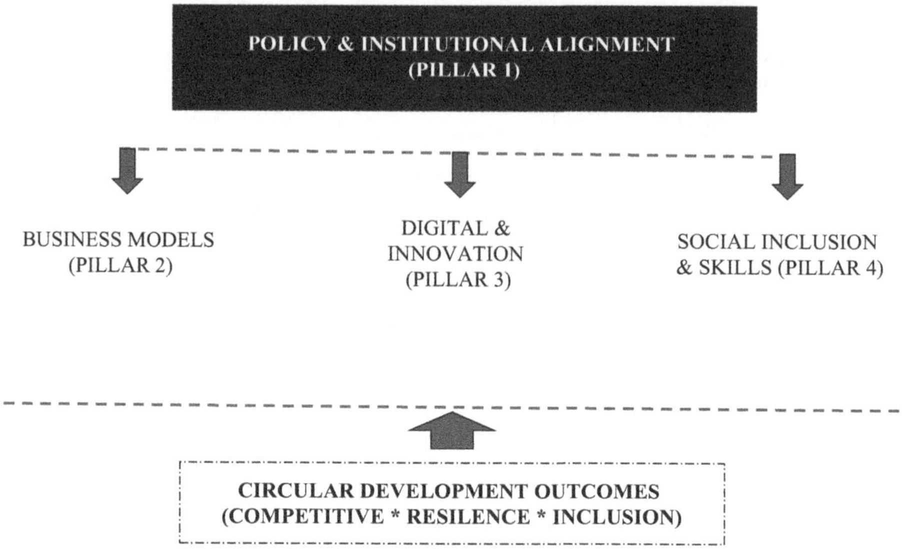

FIGURE 7.5 Integrated four-pillar framework

frameworks augment circular potential across industries and reinforce MSW systems (Godfrey et al., 2021; Sesay & Fang, 2025). In the absence of regulatory clarity, enforcement capability, and institutional coordination, circular projects are scattered and ephemeral.

Pillar 2: Circular business models and entrepreneurship convert policy aspirations into economic benefits. Small and medium-sized organizations, start-ups, and informal businesses implement reuse, repair, recycling, remanufacturing, and product–service systems. In developing countries, entrepreneurship serves as a significant catalyst for job development, resource optimization, and local value preservation (Panait et al., 2022). Collaboration throughout value chains enhances resource efficiency and facilitates industrial symbiosis (Mishra et al., 2021). Circular transitions yield significant socioeconomic advantages, such as employment generation and livelihood diversification (Clube, 2022).

Pillar 3: Digitalization, research, and innovation facilitate scalability, coordination, and adaptive learning. Technologies associated with Industry 4.0, digital waste tracking platforms, and traceability systems enhance the visibility of material movement and the efficiency of reverse logistics (Okorie et al., 2018; Cheah et al., 2022). Digital platforms can address coordination deficiencies prevalent in emerging economies (Nemilentseva et al., 2024), whereas green technological innovation improves long-term environmental performance and competitiveness (Tanveer et al., 2022). Research institutes are essential for producing locally pertinent knowledge and preventing the unthinking adoption of models from high-income settings (Um, 2025).

Pillar 4: Social inclusion and skills development guarantee the validity, resilience, and equity of cyclical transformations. In emerging economies, informal participants such as waste pickers, repairers, and micro-entrepreneurs already fulfill critical circular roles. Identifying and incorporating these stakeholders via formal systems enhances efficiency, increases material recovery rates, and secures livelihoods (Clube, 2022). Inclusive governance systems, including cooperatives, public–private–community partnerships, and participatory planning platforms, enhance trust and collective ownership (Mishra et al., 2021).

The enhancement of skills is fundamental to this pillar. Technical and vocational education in recycling technologies, digital systems, remanufacturing, and sustainable product design improves worker preparedness for burgeoning green industries. Capacity building could also encompass policymakers and local authorities to facilitate adaptive governance (Sesay & Fang, 2025). Behavioral change efforts, consumer awareness initiatives, and community-based projects such as decentralized composting or repair hubs integrate circular practices into everyday life (Um, 2025).

Ultimately, social inclusion converts the circular economy from a solely technical objective into a developmental approach that mitigates inequality, enhances community resilience, and guarantees the equitable distribution of economic benefits.

The four pillars operate as a cohesive system where legislation facilitates business innovation, digitization improves efficiency and learning, and social inclusion ensures long-term sustainability. The concept establishes the CE as both an environmental imperative and a means to achieve inclusive growth, competitiveness, and systemic resilience in developing nations.

7.3 KEY STAKEHOLDERS AND THEIR ROLES

7.3.1 Government and Public Institutions

Governments are essential in facilitating the development of a CE by influencing the policy, regulatory, and institutional framework. In developing economies, public institutions are tasked with incorporating CE principles into national development plans, industrial policies, waste management strategies, and urban governance frameworks. Explicit and comprehensive regulations encompassing waste segregation, EPR, green public procurement, and requirements for secondary materials furnish the certainty necessary for private investment and innovation. In the absence of continuous enforcement and interministerial coordination, circular projects can remain disjointed and ephemeral.

In addition to regulation, governments function as market makers and facilitators. The state can drive demand for circular products and services, especially in building, transit, and municipal services, through public procurement, infrastructure investment, and tax incentives. Local governments play a vital role in SWM, as decentralized decision-making affects waste collection, recycling systems, and community involvement. Capacity enhancement at the municipal level is crucial for converting national CE aspirations into practical results.

Governments bear a social obligation to ensure that CE transformations are inclusive and equitable. Policies must acknowledge informal labor, encourage skills enhancement, and facilitate SME involvement instead of prioritizing capital-intensive solutions that undermine livelihoods. Ultimately, public institutions serve as conveners, uniting industry, researchers, financiers, and communities via platforms that facilitate collaboration, experimentation, and learning. In emerging economies, proficient governmental leadership dictates whether the CE evolves into a revolutionary development instrument or persists as a series of disjointed initiatives.

7.3.2 Private Sector and Industry

The private sector is the principal executor and creator of CE methods in developing economies. Companies implement circular principles into practice via circular product design, sustainable production, reuse, remanufacturing, recycling, and service-oriented business models. SMEs and start-ups are vital because of their adaptability and closeness to local markets, whilst larger corporations serve a catalytic function by incorporating CE principles into supply chains and assisting smaller suppliers.

Investment decisions in the industry are significantly shaped by economic feasibility and risk assessment. Adoption rises if circular techniques diminish costs, ensure material supply, or enhance competitiveness, particularly in export-oriented sectors. Nevertheless, companies frequently encounter obstacles including restricted financial access, skill deficiencies, inadequate infrastructure, and unpredictable demand for circular products. Collaboration through value chains mitigates these obstacles by facilitating the exchange of resources, expertise, and infrastructure, hence promoting industrial symbiosis and reverse logistics systems.

The private sector additionally influences market dynamics and customer involvement. By implementing transparency, labeling, and quality assurance, companies may establish trust in recycled and remanufactured items. Collaborations with governmental bodies and research institutions facilitate the piloting and expansion of contextually suitable technology. In developing economies, the private sector's role encompasses not only profit generation but also job creation, supplier growth, and local economic integration. When coupled with supportive policies and inclusive governance, industry serves as a formidable catalyst for circular development.

7.3.3 Communities, Informal Sector, and Consumers

Communities, informal laborers, and consumers are pivotal to the social sustainability of CE transitions in developing nations. Informal players, like waste pickers, repairers, and micro-entrepreneurs, already provide essential circular tasks, particularly in material recovery and reuse. Recognizing and incorporating these entities inside formal institutions enhances efficiency, livelihoods, and environmental results. Exclusionary or too technical CE interventions may jeopardize established circular processes and provoke societal opposition.

Consumers impact the success of the CE by their behavioral decisions, such as waste segregation, engagement in reuse systems, and endorsement of circular products. Consequently, public awareness, education, and trust-building are vital elements of CE tactics. Community-driven initiatives such as decentralized composting, repair centers, and cooperative recycling programs enhance local ownership and resilience while decreasing system expenses.

Communities serve as co-creators of circular solutions, especially when bolstered by local governments, NGOs, and social entrepreneurs. Inclusive participation guarantees that CE initiatives align with local requirements, cultural customs, and economic conditions. In poor economies, where social and economic vulnerabilities are significant, the credibility of the CE relies on its capacity to enhance daily life, rather than solely focusing on environmental metrics. Empowered communities convert CE from a theoretical notion into a practical developmental approach.

7.3.4 Research Institutions and Innovation Systems

Research institutions and innovation systems furnish the foundational knowledge necessary for the implementation of a CE. Universities, research institutions, and innovation centers produce evidence regarding material flows, environmental impacts, business models, and policy efficacy. In developing nations, context-specific research is crucial for tailoring CE principles to distinguish socioeconomic and institutional frameworks instead of merely duplicating models from affluent economies.

Innovation systems facilitate experimentation and expansion, allowing for experimental initiatives, technological evaluation, and business model advancement. Collaboration among researchers, industry, and government expedites knowledge acquisition and mitigates the dangers linked to novel circular practices. Research institutions are essential in cultivating capabilities by teaching professionals, entrepreneurs, and policymakers in CE competencies.

In addition to technology, research fosters governance innovation by informing adaptive policies and inclusive transition strategies. Research institutions integrate science, policy, and practice to guarantee that CE policies are grounded in evidence, adaptable, and socially attuned. In developing countries, continuous investment in research and innovation is crucial for the long-term viability and resilience of circular transitions.

REFERENCES

Cheah, C.G., Chia, W.Y., Lai, S.F., Chew, K.W., Chia, S.R. and Show, P.L. (2022) 'Innovation designs of Industry 4.0 based solid waste management: Machinery and digital circular economy', *Environmental Research*, 213, p. 113619. https://doi.org/10.1016/j.envres.2022.113619

Clube, R. (2022) 'Is job creation a legitimate social benefit of the circular economy?' *Resources, Conservation and Recycling*, 181, p. 106220. https://doi.org/10.1016/j.resconrec.2022.106220

Godfrey, L.K., Nahman, A., Oelofse, S.H., Trotter, D., Khan, S., Nontso, Z., Magweregwede, F., Sereme, B.V., Okole, B.N., Gordon, G.E. and Brown, B. (2021) 'The CEas development opportunity: Exploring CEopportunities across South Africa's economic sectors', *South African Journal of Science*, 117(3/4), pp. 1–11. https://doi.org/10.17159/sajs.2021/8169

Mishra, J.L., Chiwenga, K.D. and Ali, K. (2021) 'Collaboration as an enabler for circular economy: A case study of a developing country', *Management Decision*, 59(8), pp. 1784–1800. https://doi.org/10.1108/MD-10-2018-1111

Nemilentseva, M., Tariq, A., Tariq, W., Aghajani, D. and Torkkeli, M. (2024) 'A review of digital platform and circular economy: Opportunities and challenges for developing countries', in Umair, S., Waqas, U., Mrugalska, B. and Al Shamsi, I.R. (eds.) *Human Perspectives of Industry 4.0 Organizations: Reviewing Sustainable Performance*. Boca Raton, FL: CRC Press, pp. 38–67. https://doi.org/10.1201/9781032616810-3

Okorie, O., Salonitis, K., Charnley, F., Moreno, M., Turner, C. and Tiwari, A. (2018) 'Digitisation and the circular economy: A review of current research and future trends', *Energies*, 11(11), p. 3009. https://doi.org/10.3390/en11113009

Panait, M., Hysa, E., Raimi, L., Kruja, A. and Rodriguez, A. (2022) 'Guest editorial: CEand entrepreneurship in emerging economies: Opportunities and challenges', *Journal of Entrepreneurship in Emerging Economies*, 14(5), pp. 673–677. https://doi.org/10.1108/JEEE-10-2022-0487

Scheel, C. and Aguiñaga, E. (2025) 'A systems view of circular economy', *Sustainability*, 17(3), p. 1268. https://doi.org/10.3390/su17031268

Sesay, R.E.V. and Fang, P. (2025) 'CEin municipal solid waste management: Innovations and challenges for urban sustainability', *Journal of Environmental Protection*, 16(2), pp. 35–65.

Tanveer, M., Khan, S.A.R., Umar, M., Yu, Z., Sajid, M.J. and Haq, I.U. (2022) 'Waste management and green technology: Future trends in CE leading towards environmental sustainability', *Environmental Science and Pollution Research*, 29(53), pp. 80161–80178. https://doi.org/10.1007/s11356-022-23238-8

Um, N. (2025) 'Sustainable waste management in the context of the circular economy', *Sustainability*, 17(5), p. 1937. https://doi.org/10.3390/su17051937

8 Conclusion

8.1 OVERVIEW

This book presents a logical and contextually aware interpretation of the CE as a developmental trajectory and a systems transformation initiative, specifically emphasizing SWM in developing nations. It transcends mere conceptual conceptions of the CE, exploring the operationalization, governance, financing, and social integration of CE principles within contexts marked by increasing urbanization, informality, institutional limitations, and resource scarcity.

This book's primary intellectual contribution is its assertion that solid waste should be viewed not just as an environmental liability but as a strategic resource that can provide economic value, employment, and social inclusion when managed through circular systems. It reconceptualizes waste management from a mere municipal duty to a multifaceted development challenge interconnected with industrial policy, supplier chains, innovation systems, labor markets, and social equality.

This book presents the CE as a mechanism for attaining comprehensive sustainability objectives, such as enhanced public health, diminished environmental degradation, climatic resilience, equitable growth, and the fortification of institutions. It notably avoids the unthinking application of CE models from rich economies, instead of highlighting the importance of contextual adaptation, gradual transformation, and locally rooted solutions.

8.2 CORE INSIGHTS ON CE THINKING

A consistent theme throughout this book is that the CE is not a singular model, but rather a collection of techniques that must be harmonized with local economic frameworks, governance capabilities, and social contexts. This book emphasizes that the implementation of CE in poor nations is frequently hindered not by a deficiency of ideas but by disjointed institutions, inadequate coordination, and a misalignment between policy objectives and practical execution.

This book presents numerous fundamental concepts:

- The CE necessitates a systems approach that incorporates production, consumption, waste generation, recovery, and reintegration into value networks.
- SWM serves as a pivotal entry point for the CE due to its visibility, labor intensity, and strong connection to urban livelihoods.
- Informality, frequently regarded as an obstacle, can be utilized as a transitional asset if well incorporated, governed, and endorsed.

DOI: 10.1201/9781003591764-8

- Digitalization, innovation, and data serve as enablers rather than solutions, necessitating institutional preparedness and human capability to generate value.
- Social inclusion and skills development are the fundamental components of the success of a CE.

These findings collectively contest linear, end-of-pipe waste management methods and advocate for preventive, regenerative, and value-retentive alternatives.

8.3 SIGNIFICANCE OF CE IN SWM

This book's significant contribution is in its thorough examination of the necessity of a CE for contemporary SWM, particularly in swiftly urbanizing areas.

8.3.1 Moving Beyond Collection and Disposal

Conventional SWM methods in numerous developing nations are predominantly focused on collection and disposal, frequently depending on open dumping or inadequately managed landfills. This book illustrates that these systems are economically unviable, ecologically detrimental, and socially restrictive.

The CE instigates a fundamental transformation by emphasizing:

- Prevention and reduction of waste at the source.
- Separation and sorting to maintain material value.
- Recovery, recycling, composting, and reuse.
- Reintegration of materials into industrial systems.

This transition redefines waste management from a cost center to a value-generating system, capable of bolstering local companies, diminishing import reliance, and prolonging product lifespans.

8.3.2 Waste as a Resource and Economic Opportunity

A major finding highlighted in this book is that solid waste streams plastics, organics, paper, metals, e-waste, and building debris embody untapped economic value. When regulated by circular frameworks, these streams can facilitate recycling and reprocessing firms:

- Composting and soil restoration in agriculture
- Waste-to-value small and medium-sized enterprises and cooperatives
- Green employment in collecting, processing, and manufacturing

This book emphasizes that CE-based waste systems are especially conducive to employment development, as they require more labor than landfill-based systems. This is particularly important in environments characterized by elevated youth unemployment and informal labor marketplaces.

8.3.3 Governance, Policy, and Institutional Alignment

A key finding from this book is that the CE in SWM necessitates robust governance alignment for success. This book emphasizes that fragmented responsibilities among ministries, local authorities, and agencies frequently hinder CE projects.

Essential governance insights highlight the necessity for policy coherence among waste management, industrial development, environmental regulation, and local government institutions:

- The significance of EPR as a strategy for reallocating responsibility upstream and funding recovery mechanisms.
- The function of municipalities extends beyond mere service provision to encompass their role as system orchestrators that facilitate collaborations with private sector entities, communities, and informal laborers.
- The imperative for regulatory frameworks that promote innovation while ensuring protections for environmental and social results.

This book presents governance reform as a facilitator of circularity, rather than a bureaucratic hindrance, and underscores the importance of flexible, learning-oriented organizations.

8.3.4 Business Models, Entrepreneurship, and Supply Chains

This book convincingly argues that the effectiveness of solid waste circularity is fundamentally achieved through sustainable business strategies and robust supply chains. This examines the adaptation of circular business models such as recycling-based manufacturing, product–service systems, reverse logistics, and shared infrastructure to contexts in emerging economies.

Critical insights reveal that circular firms frequently encounter challenges not owing to insufficient demand, but rather because of funding deficiencies, irregular material supply, and inadequate market connections:

- Incorporating waste pickers and informal collectors into organized value chains enhances material recovery rates and concurrently improves livelihoods.
- Public procurement and sustainable purchasing can serve as a catalyst for establishing stable markets for recycled and circular products.
- Integration of the supply chain is crucial for transitioning from pilot initiatives to scalable CE systems.

This book recontextualizes SWM within the framework of comprehensive circular supply chain ecosystems, rather than as discrete municipal services.

8.3.5 Digitalization, Data, and Innovation in Waste Systems

This book emphasizes the facilitative impact of digitalization and innovation in enhancing circular SWM across multiple parts. Nonetheless, it maintains a balanced perspective, warning against technology-centric solutions that overlook institutional and social realities.

Principal observations encompass the following:

- Digital tools enhance traceability, oversight, and efficiency in waste collection and recycling processes.
- Data is essential for planning, investment decisions, and performance evaluation in CE systems.
- Innovation in developing economies is frequently characterized by frugality and adaptability, arising from limitations rather than excess.
- Collaboration among research, industry, and government is crucial for the localization of technology and business models.

This book underscores that innovation ecosystems require intentional cultivation, with universities and research organizations serving as information brokers.

8.3.6 Social Inclusion, Informality, and Skills Development

This book's notable contribution is its persistent emphasis on social inclusion. Instead of viewing informal waste actors as a challenge to be eradicated, this book acknowledges them as crucial agents of circularity.

Key insights encompass the following:

- Informal waste pickers achieve significant recovery rates and environmental advantages, despite operating under precarious circumstances.
- The CE offers a framework for formalization that is inclusive, utilizing cooperatives, contracts, training, and social protection.
- Skills development is crucial across various levels, ranging from fundamental occupational safety to entrepreneurship, digital literacy, and management.
- The gender aspects of waste work necessitate explicit focus to guarantee equitable participation and benefit distribution.

By emphasizing social inclusion, this book frames CE-based SWM as a strategy for equitable transition rather than solely an environmental measure.

8.4 IMPLICATIONS FOR SUSTAINABLE DEVELOPMENT

This book regularly ties circular SWM with overarching development objectives. It illustrates how CE enhances:

- Environmental sustainability via diminished pollution and resource extraction.
- Economic resilience through local value generation and SME development.
- Social outcomes through job creation, health advancements, and inclusivity.
- Climate action through lowered emissions and enhanced resource efficiency.

This book contends that CE in SWM provides co-benefits, rendering it an effective and practical avenue for sustainable development in resource-limited contexts.

8.5 FORWARD-LOOKING REFLECTIONS AND STRATEGIC DIRECTION

This book underscores that the shift to circular SWM is evolutionary, not instantaneous. It necessitates a long-term vision grounded in pragmatic implementation strategies:

- Experiential learning via scalable pilot programs
- Enhanced incorporation of CE into national development strategies
- Investment in institutions, competencies, and data infrastructures

This book asserts that a CE, rooted in local contexts and propelled by participatory governance, has the potential to revolutionize SWM, transforming a significant urban burden into an opportunity for sustainable change.

This CE Book offers a substantial, interdisciplinary, and application-focused addition to the literature on CE. Its focus on SWM is both current and purposeful, acknowledging waste as a concrete foundation for wider circular transitions.

The primary lesson conveyed in this book is unequivocal: SWM is not a marginal environmental concern but a fundamental developmental challenge and opportunity. The CE provides conceptual tools, legislative frameworks, and practical tactics to convert waste systems into catalysts for equitable and sustainable growth.

BIBLIOGRAPHY

Ellen MacArthur Foundation (2013) *Towards the Circular Economy: Economic and Business Rationale for an Accelerated Transition*. Cowes: Ellen MacArthur Foundation.

Elia, V., Gnoni, M.G. and Tornese, F. (2020) 'Measuring circular economy strategies through supply chains', *Business Strategy and the Environment*, 29(2), pp. 547–560.

Geng, Y., Sarkis, J. and Ulgiati, S. (2016) 'Sustainability, well-being, and the circular economy in developing countries', *Journal of Cleaner Production*, 133, pp. 18–29.

Geissdoerfer, M., Savaget, P., Bocken, N.M.P. and Hultink, E.J. (2017) 'The circular economy—A new sustainability paradigm?' *Journal of Cleaner Production*, 143, pp. 757–768.

Kirchherr, J., Reike, D. and Hekkert, M. (2017) 'Conceptualizing the circular economy: An analysis of 114 definitions', *Resources, Conservation and Recycling*, 127, pp. 221–232.

Mwanza, B.G. and Mbohwa, C. (2019) 'Circular economy and solid waste management in developing countries', *Procedia Manufacturing*, 33, pp. 91–98.

Paes, L.A.B., Bezerra, B.S. and Godoy, M.L. (2024) 'Circular waste management as a development pathway in emerging economies', *Journal of Cleaner Production*, 431, p. 140123.

Scheel, C., Aguiñaga, E. and Bello, B. (2020) 'Decoupling economic development from resource consumption: A systems thinking approach to circular economy', *Journal of Cleaner Production*, 268, p. 122111.

Zaman, A.U. and Lehmann, S. (2011) 'Challenges and opportunities in transforming cities into zero-waste cities', *Challenges*, 2(4), pp. 73–93.

Index

Note: *Italic* page numbers refer to figures.

A

African Development Bank (AfDB), 70, 72
Agricultural and organic waste management, 107
  circular practices, 108–109
    application, 109–110
    bioenergy recovery through AD and biogas production, 110–111
    urban systems and design, 109
    zero-waste farming system, 111
  importance, 107–108
Ahmedabad model, 94–95
Anaerobic digestion (AD), 17, 43, 44, 46, 47, 49, 51, 73, 107–109, 127
  bioenergy recovery through, 110–111
  projects in Ethiopia and South Africa, 47
  segregation, 42
  waste treatment technologies, 44
Artificial intelligence (AI), 14, 73–75, 126

B

Bangladesh, CE model of developing countries from, 119–120
Behavioral change, 16, 49, 50, 80, 84, 85, 91, 126, 127, 132, 137
  campaigns, 78
  communication, 82–83
Biodegradable waste, 27, *27*, 41, 77, 107
Bioenergy recovery, through AD and biogas production, 110–111
Biogas production, 16, 30
  bioenergy recovery through, 110–111
  to capture methane, 56–57
Bioreactor landfills, 57
Block collection points, in Kitwe, 32
Business initiatives, 71–72

C

Centralized sorting, and recycling facilities, 41
Changing behaviors strategies, 82–84
  behavior change communication and social awareness campaigns, 82–83
  circular economy
    education and structured learning platforms, 82
    school curricula and youth empowerment programs, 83–84
  community-based motivational mechanisms, 83
  e-learning modules, 84
  multi-stakeholder collaboration, 84
  social norms, 83
  user-friendly systems, 83–84
Chelstone Township, household participation in, 85
Circular business models, 1, 50, 66, 68, 70, 74, 84, 93, 123, 128–130, 132–134, *134*, 136, 143
  circular economy, 117–118
  in India, 117–118
Circular economy (CE), 141–142
  business models in India, 117–118
  CE-based industrial waste practices in Nigeria, 106
  challenges
    Indian SMEs, 123–124
    South Africa economy, 121–122
    stakeholders' perspectives, 122–123
    sub-Saharan Africa, 120–121
  circular industrial symbiosis, 65
  community engagement and awareness
    changing behaviors strategies, 82–84
    community recycling, demonstration, and innovation hubs, 78–79
    influencing public participation. *See* public participation strategies
    informal waste pickers, 79
    multi-stakeholder collaboration, 79
    participatory waste governance, 77–78
    school-based and youth-focused CE education, 78
    systemic and community-centered integrated waste management, 79
    tailored awareness and behaviour change campaigns, 78
  community recycling, demonstration, and innovation hubs, 78–79
  definition, 1–2
  Deposit Return Schemes, 66–67
  of developing countries from Bangladesh, 119–120
  developing economies, 11–12
    demographic, 12–13
    infrastructure deficiencies, 14–15
    institutional quality and governance, 13–14

social and environmental dimensions, 15
structure, 12
extended producer responsibility, 64–65
financial and economic tools, 69
economic incentives for consumers and businesses, 71–72
investments and funding mechanisms, 70–71
four-pillar framework
circular business models and entrepreneurship, 133–134, *134*
digitalization, research, and innovation, 134, *135*
integration, 135–137, *136*
policy and institutional alignment pillar, 133, *133*
social inclusion and skills development, 134–135, *135*
framing, 92
Namibia's mining sector, 116–117
policy and regulatory frameworks
in developing economies, 62
extended producer responsibility, 62–63
municipal solid waste ordinances and bylaws, 63
national waste management strategies and acts, 63–64
plastic bag bans and levies, 63
policy recommendations for, 68–69
practices, 104–105, 108–109
application, 109–110
bioenergy recovery through AD and biogas production, 110–111
circular industrial engineering, 104
circular supply chains and collaborative networks, 105
handling of hazardous industrial waste, 103–104
integration, 99
knowledge, attitudes, and organizational culture, 105
resource recovery and material looping, 104
sustainable operational strategies, 105
urban systems and design, 109
waste minimization through cleaner production and process reengineering, 104
zero-waste farming system, 111
principles in industrial solid waste management, 118–119
product design and innovation, 65–66
public–private partnerships, 67–68
recycling, 49–51, *50*
research and innovation, role of, 130
in school curricula, 83–84
solid waste management, 126
behavioral change, 127
business models, 143
collaboration and networks, 129
collection and disposal, 142
developing economies, 15–18
digitalization and innovation, 143–144
digital platforms and technology-enabled circular opportunities, 128
digital technologies and Industry 4.0, 126–127
economic opportunities and job creation, 3–4
employment development and social inclusion, 129–130
entrepreneurship and innovation as engines, 128
environmental sustainability and climate change mitigation, 4
green technologies, 127
implementation, 145
implications for sustainable development, 144–145
informal sector, 4
integrated and adaptive circular waste systems, 128
integrated SWM, 5
integration in urban waste governance, 127
resource and economic opportunity, 142
robust governance, 143
sector-specific opportunities, 128–129
social inclusion, 144
systems thinking and integrated circular transitions, 129
waste reduction and resource efficiency, 3
stakeholders, 137–139
communities, informal laborers, and consumers, 138–139
government and public institutions, 137–138
private sector and industry, 138
research institutions and innovation systems, 139
stakeholders' perspectives, 122–123
strategies, 93
for organic and plastic waste, 100
technological innovation, 72–73
adaptation for developing economies, 76–77
digital solutions for SWM and tracking, 74–75
emerging technologies in SWM, 73–74
urban solid waste management
Ahmedabad model, 94–95
community-based waste innovation in Mombasa County, 96
India, 94–95
innovation, 94–95

Kawasaki city in Japan, 94
smart city circular waste innovations in Nigeria, 95
zero-waste cities and municipal programs, 67
Circular industrial symbiosis, 65
Circular urban waste system, with Sustainable Development Goals (SDGs), 96
Circular value chain, 81, 104, 128
Cluster-oriented strategies, 132
Coherent laws, 135
"Collect–transport–dump" approach, 8, 79, 91
Community-based circular waste management, in Tan Deaw sub-district, 84–85
Community-based enterprises (CBEs), 85
Community-based management, 100
Community-based model, 33–34
Community-based motivational mechanisms, 83
Community bin systems, 32
Community-centered CE education, 82
Community-centered integrated waste management, 79
Community-driven initiatives, 138–139
Community engagement and awareness strategies
changing behaviors strategies, 82–84
community recycling, demonstration, and innovation hubs, 78–79
influencing public participation. *See* public participation strategies
informal waste pickers, 79
multi-stakeholder collaboration, 79
participatory waste governance, 77–78
school-based and youth-focused CE education, 78
systemic and community-centered integrated waste management, 79
tailored awareness and behaviour change campaigns, 78
Composite waste, 29, *30*
Consumer awareness, economic incentives in, 71–72
Consumers impact, 138–139
Critical resources, recovery of, 51–52
Curbside collection, 31–32

D

Decentralized and low-cost waste systems, 99–100
Decentralized/community-based model, 33–34
Decentralized composting, in India, 46–47
Decentralized transport model, 33–34
Decision support system, for iron and steel sector of Southern Africa, 106
DEMATEL methodology, 92
Deposit Return Schemes (DRS), 66–69, 72
Development Bank of Southern Africa (DBSA), 70, 72
Digital and integrated technologies, 45–46
Digitalization, research, and innovation pillar, 134, *135*, 136
Digital platforms, 68, 84, 93, 95, 127, 128, 130, 132, 134, 136
Digital solutions, for SWM and tracking, 74–75
Digital technologies, 46, 48–49, 65, 74, 75, 92–93, 126–127
Direct haul model, 33, 37
Disposal techniques
illegal disposal solid waste, 57–58
open burning, 55–56
open dumping, 55
Door-to-door collection/systems, 31–32
and recycling, 38
Drop-off systems, 31–32
Dynamic microbial ecosystems, 56–57

E

Education and behavioral modification programs, 101
Eighth National Development Plan (8NDP), 70, 73
Emerging technologies, in SWM, 73–74
Employment development, 129–130, 142
"End-of-life" concept, 1, 2, 64, 68
Entrepreneurship
of CE growth, 128
pillars of, 133–134, *134*
Environmental, and public health impacts, 99
Equity, in waste systems, 81
Ethiopia, AD projects in, 47
Extended producer responsibility (EPR), 3, 10, 41, 62–65, 105, 133, 135, 137, 143

F

Financial resources, 5, 7, 33, 37, 45, 72, 91
and economic tools, 69
economic incentives for consumers and businesses, 71–72
investments and funding mechanisms, 70–71
limitations, 18
Four-pillar framework, CE
circular business models and entrepreneurship, 133–134, *134*
digitalization, research, and innovation, 134, *135*
integration, 135–137, *136*
policy and institutional alignment pillar, 133, *133*
social inclusion and skills development, 134–135, *135*
Funding mechanisms, 70–71

G

Garbage collection services, 90, 96
Geographic Information Systems (GIS), 74
  smart planning with, 100–101
Global Environment Facility (GEF), 70
Government, and public institutions, 137–138
Green Climate Fund (GCF), 70
Green Economy Strategy and Implementation Plan (GESIP), 70
Green Growth Strategy, 70
Greenhouse gas (GHG) emissions, 4, 5, 7, 9, 11, 16, 18–20, 38, 47, 49, 55, 57, 65, 73, 92, 102, 107–110, 127
Green technologies, 124, 127

H

Hazardous waste, 17, 27–29, *29*, 40, 65, 103–105
High waste generation, 16–17
Home Appliance Recycling Law, 65
Household participation, in Chelstone Township (Lusaka), 85
Hybrid recycling systems, Johannesburg of, 48

I

Illegal disposal solid waste, 57–58
Inclusive models, 127
India
  CE innovation, 94–95
  circular business models in, 117–118
  decentralized composting in, 46–47
  mechanical–biological treatment plants, 47–48
  mixed municipal waste, 47–48
  waste-to-energy technologies, 47–48
Indian SMEs, 123–124
Industrial ecology (IE), 65
Industrial solid waste management, 102–103
  case studies
    CE-based industrial waste practices in Nigeria, 106
    decision support system for iron and steel sector of Southern Africa, 106
    Lebanon's industrial waste management, 105–106
    synthesis, 106
  CE practices, 104–105
    circular industrial engineering, 104
    circular supply chains and collaborative networks, 105
    handling of hazardous industrial waste, 103–104
    knowledge, attitudes, and organizational culture, 105
    minimization through cleaner production and process reengineering, 104
    resource recovery and material looping, 104
    sustainable operational strategies, 105
  circular economy principles, 118–119
Industrial symbiosis (IS), 65, 70, 92, 102, 104, 106, 117, 119, 129, 136, 138
Industry 4.0-enabled CE systems, 126–127
Industry 4.0 pilot programs, 48–49
Industry 4.0 technologies, 132, 134, 136
Inert waste, 27, *28*
Informal integration, of Johannesburg, 48
Informal sector, 37–38, 138–139
  challenges, 39
  collection, 32
  door-to-door collection and recycling, 38
  environmental impact, 38–39
  formalization and integration, 39–40
  involvement, 17
  livelihood and socioeconomic, 38
  model and private sector entities, 34
Informal waste pickers, 16, 19, 20, 32, 34, 38, 40, 45, 48, 55, 62, 63, 71, 74, 76, 79, 81, 95, 121, 144
Innovation-driven growth, 128
Innovation systems, 132, 139, 141
Integrated SWM (ISWM), 5
Intermodal model, 34–35, 37
Internet of Things (IoT), 48, 73–75, 92, 126

J

Job creation, 3–4, 49, 56, 69, 76, 117, 119, 129–130, 132, 138

K

Kawasaki city, in Japan, 94
Kenya, 12, 14, 62–64, 69–71, 73, 76, 77, 91
  community-based waste innovation in Mombasa County, 96

L

Landfilling techniques
  engineering and design features, 56
  leachate quality and treatment, 56–57
Leachate quality and treatment, 56–57
Lebanon's industrial waste management, 105–106
Life cycle assessments (LCAs), 47
Local leadership, 67, 69, 79, 82, 101
Local ownership, of CE, 80
Lusaka, 26, 31, 63, 85
  household participation in Chelstone Township, 85

M

Material recovery rates, 70, 71, 81, 137, 143
Materials recovery facilities (MRFs), 41, 46, 96
  hybrid recycling systems, 48
  recycling and, 45
Mechanical–biological treatment (MBT) plants, in India, 47–48
Mediocre waste management, socioeconomic and environmental effects, 18–20
Micro, small, and medium-sized enterprises (MSWMs), 73
Mixed municipal waste, in India, 47–48
Mobile technology, 128
Mombasa County, community-based waste innovation in, 96
M-Pesa, 12, 14
Multi-stakeholder circular networks, 129
Multi-stakeholder collaboration, 79, 81, 102
  for coordinated messaging, 84
Municipal programs, 67
Municipal solid waste (MSW), 4, 6, 7, 9, 16, 17, 26, 27, 30, 31, 37, 41–43, 46, 62, 63, 67, 92–94, 136
  ordinances and bylaws, 63

N

Namibia's mining sector, 116–117
National Environmental Management, 62, 63
National waste management strategies, 63–64
New product development (NPD), 66
Nigeria
  CE-based industrial waste practices in, 106
  smart city circular waste innovations in, 95
Nonbiodegradable waste, 90, 98
Nongovernmental organizations (NGOs), 41, 48, 79, 84, 138
Northern Greece, reward recycling and pay-as-you-throw, 85

O

Online learning, 84
Open burning technique, 55–56
Open dumping technique, 55
Organic waste, dominance of, 17
Original equipment manufacturers (OEMs), 122

P

Paris Agreement, 4
Participatory action research (PAR), 84
Participatory waste governance, 77–78
Pay-as-you-throw (PAYT), in Northern Greece, 85
Plastic bag charge, 63
Plastic recycling processes, 51
Plastic solid waste (PSW), 49
Policy and institutional alignment pillar, 133, *133*, 135–136
Poor community participation, 91
Poor waste segregation, 90
Private sector
  and industry, 138
  entities and informal model, 34
Public health risks, SWM, 18
Public participation strategies, 80–82
  behavioral institutionalization, 82
  knowledge, awareness, and behavioral change, 80
  material recovery rates, 81
  multi-stakeholder collaboration, 81
  ownership and acceptance of CE, 80
  policy effectiveness and service delivery, 80–81
  social inclusion, 81
Public–private partnerships (PPPs), 67–68, 96

R

Rapid urbanization, 16, 19, 29, 37, 90
Recyclable waste, 27, *28*
Recycled PET (rPET), 51, 54
Recycling
  circular economy, 49–51, *50*
  and door-to-door collection, 38
  materials, 53
    barriers and market dynamics, 54
    demand and applications, 54
    market growth, 54
    market innovation and supply chains, 54–55
  and MRFs, 45
  practices, 90
  systems, 51–53, *53*
    life cycle perspectives, 52–53
    plastic recycling processes, 51
    recovery of critical resources, 51–52
Refuse-derived fuel (RDF), 44, 48, 93
Regenerative business models, 93
Remote sensing, smart planning with, 100–101
Research institutions, 116, 130, 138, 139
Resource cycles, recovery, reuse, and recycling of, 104
Resource recovery innovations, 127
Reward recycling (RR) systems, in Northern Greece, 85
Robotics, 73
Robust governance, 11, 44, 109, 143
Rural solid waste management, 96–97
  challenges, 97–98
    environmental and public health impacts, 99
    infrastructure and services, 98

integration of CE practices, 99
low public awareness and cultural practices, 98
national budgets and donor resources, 98
nonbiodegradable waste, 98
qualified personnel, 98–99
weak policy enforcement, 99
program, 101–102
tailored solutions
CE strategies for organic and plastic waste, 100
community-based management, 100
decentralized and low-cost waste systems, 99–100
education and behavioral modification programs, 101
remote sensing and GIS, 100–101

S

School-based CE education, 78
Sector-specific opportunities, 128–129
Sector-specific strategies, 132
Small and medium enterprises (SMEs), 117, 119, 122, 123
CE principles in industrial, 118–119
Indian SMEs, 123–124
Smart city circular waste innovations, in Nigeria, 95
Smart waste management technologies, 92–93
Smart Zambia Master Plan, 75
Social awareness campaigns, 82–83
Social inclusion, 39, 41, 46, 72, 74, 76, 81, 93–95, 97, 127, 129–130, 132–137, 141, 142, 144
and skills development, 134–137, *135*
Social norms, 83, 101, 123
Solid waste
case studies
AD projects in Ethiopia and South Africa, 47
decentralized composting in India, 46–47
digital technologies/Industry 4.0 pilot programs, 48–49
Johannesburg of hybrid recycling systems, 48
mechanical–biological treatment plants, 47–48
challenges, 5–6
emerging and complex waste streams, 7–8, 10
environmental and public health impacts, 7
inequalities, 9
municipal solid waste, 6
public health hazard, 9–10
socioeconomic dimensions, 8
Sustainable Development Goal, 10–11
sustainable practices, 8–9
tackling socioeconomic vulnerabilities, 10
volumes, 9
waste management capacity, 7
characteristics, 29–31
disposal techniques
illegal disposal solid waste, 57–58
open burning, 55–56
open dumping, 55
landfilling techniques
engineering and design features, 56
leachate quality and treatment, 56–57
recycling
in CE, 49–51
materials, 53–55
systems, 51–53
segregation
anaerobic digestion, 42
centralized sorting and recycling facilities, 41
composting, 41–42
informal sector sorting, 40–41
source segregation, 40
waste-to-energy technologies, 42–43
sources, 26
transportation challenges
environmental and health concerns, 36
high operational costs, 35
infrastructure limitations, 35
institutional and financial barriers, 36
rapid urbanization, 37
transfer stations, 35–36
weak integration of informal sector, 36
transportation model
decentralized/community-based model, 33–34
direct haul model, 33
intermodal model, 34–35
private sector entities and informal model, 34
station or transfer-based model, 33
types, 26–29
waste collection systems, 31–33, *See also* waste collection systems
waste treatment technologies
anaerobic digestion, 44
comparative analysis, 46
composting, 43–44
digital and integrated technologies, 45–46
recycling and MRFs, 45
waste-to-energy technologies, 44–45
Solid waste management (SWM), 62, 64, 70, 72, 76, 77, 80, 82, 84, 85, 126, 133, 135, 137
Act of 2015 in Nairobi, 63

agricultural and organic waste. *See* agricultural and organic waste management
behavioral change, 127
business models, 143
collaboration and networks, 129
collection and disposal, 142
developing economies, 15–18
digitalization and innovation, 143–144
digital platforms and technology-enabled circular opportunities, 128
digital solutions for, 74–75
digital technologies and Industry 4.0, 126–127
economic opportunities and job creation, 3–4
emerging technologies, 73–74
employment development and social inclusion, 129–130
entrepreneurship and innovation as engines, 128
environmental sustainability and climate change mitigation, 4
green technologies, 127
implementation, 145
implications for sustainable development, 144–145
industrial SWM. *See* industrial solid waste management
informal sector, 4
integrated and adaptive circular waste systems, 128
integrated SWM, 5
integration in urban waste governance, 127
resource and economic opportunity, 142
robust governance, 143
rural SWM. *See* rural solid waste management
sector-specific opportunities, 128–129
social inclusion, 144
systems thinking and integrated circular transitions, 129
urban SWM. *See* urban solid waste management
waste reduction and resource efficiency, 3
Solid Waste Regulation and Management Act, 64
South Africa, 38–45, 63, 69–71, 73, 76, 77
AD projects in, 47
economy, 121–122
hybrid recycling systems with MRFs, 48
Southeast Asia, digital/Industry 4.0 pilot programs, 48–49
Southern Africa, decision support system for iron and steel sector of, 106
Specialized collection systems, 32–33
Stakeholders, CE, 137–139
communities, informal laborers, and consumers, 138–139
government and public institutions, 137–138
perspectives, 122–123
private sector and industry, 138
research institutions and innovation systems, 139
Station/transfer-based model, 33
Structured learning platforms, 82
Sub-Saharan Africa, 15, 16, 120–121
digital/Industry 4.0 pilot programs, 48–49
Sustainable Development Goals (SDGs), 4–5, 10–11, 39, 67, 126
circular urban waste system with, 96
Sustainable development, implications for, 144–145
Sustainable Waste Management Act, 62, 64
of Kenya, 71
Systems-thinking perspective, 79, 129

T

Tailored awareness, 78
Tailored solutions
rural solid waste management
CE strategies for organic and plastic waste, 100
community-based management, 100
decentralized and low-cost waste systems, 99–100
education and behavioral modification programs, 101
remote sensing and GIS, 100–101
Tan Deaw, sub-district in Thailand, 84–85
Thailand, Tan Deaw sub-district in, 84–85
Tracking, digital solutions for, 74–75
Transfer-based model, 33
Transportation model
challenges
environmental and health concerns, 36
high operational costs, 35
infrastructure limitations, 35
institutional and financial barriers, 36
rapid urbanization, 37
transfer stations, 35–36
weak integration of informal sector, 36
decentralized/community-based model, 33–34
direct haul model, 33
intermodal model, 34–35
private sector entities and informal model, 34
station or transfer-based model, 33

U

Urban solid waste management
challenges, 90
deficiency of environmental protections, 91
financial resources, 91
governance, policy gaps, and enforcement challenges, 91
inadequate garbage collection services, 90

low public awareness, 91
poor waste segregation, 90
rapid urban population, 90
circular economy
Ahmedabad model, 94–95
community-based waste innovation in Mombasa County, 96
India, 94–95
innovation, 94–95
Kawasaki city in Japan, 94
smart city circular waste innovations in Nigeria, 95
innovations, 91
CE framing, 92
climate change, 92
policies and social inclusion, 93–94
regenerative business models, 93
smart city and CE strategies, 93
smart waste management technologies, 92–93
Urban symbiosis model, 94
Urban waste governance, CE integration in, 127
User-friendly systems, 83–84

W

Waste Act, 63–64
Waste collection systems
block collection points in Kitwe, 32
curbside and door-to-door systems, 31–32
drop-off systems, 31–32
informal sector, 37–38
challenges, 39
collection, 32
door-to-door collection and recycling, 38
environmental impact, 38–39
formalization and integration, 39–40
livelihood and socioeconomic, 38
specialized collection systems, 32–33
Waste reduction and resource efficiency, 3
Waste streams, as input resources, 93
Waste-to-energy (WtE) technologies, 44–45
India, 47–48
segregation, 42–43
Waste treatment technologies
anaerobic digestion, 44
comparative analysis, 46
composting, 43–44
digital and integrated technologies, 45–46
recycling and MRFs, 45
waste-to-energy technologies, 44–45
Weak institutional capacity, 91
Weak policy enforcement, 99

Y

Youth empowerment programs, 83–84
Youth-focused CE education, 78

Z

Zambia, household participation in Chelstone Township, 85
Zero-waste cities, 67, 69
Zero-waste farming system, 111

For Product Safety Concerns and Information please contact our EU representative GPSR@taylorandfrancis.com Taylor & Francis Verlag GmbH, Kaufingerstraße 24, 80331 München, Germany